MEDICAL GENETICS

WIDUKIND LENZ

MEDICAL GENETICS

TRANSLATED BY ELISABETH F. LANZL

THE UNIVERSITY OF CHICAGO PRESS

This book was originally published as

Medizinische Genetik: Eine Einführung in ihre Grundlagen und Probleme

by Georg Thieme Verlag, Stuttgart, 1961

Library of Congress Catalog Card Number: 63-11399

The University of Chicago Press, Chicago & London

The University of Toronto Press, Toronto 5, Canada

PREFACE

During the past few years, research on human heredity, or human genetics, has drawn new methods and newly posed questions into the scope of its investigations. In this process, advances have been made which do not consist simply of an amassing of new knowledge on a firmly established basis. In many instances, concepts which had been carried down from past investigations have had to be critically examined. It has been possible in part to solve, and in part to formulate them more clearly; in addition, new problems have evolved. Hardly a year passes in the course of which the collaboration between human genetics and its related disciplines does not reveal undreamed-of new prospects. The physician has begun to recognize more and more that human genetics is not a study of remote peculiarities, but that it leads right into the midst of medical problems of current interest. An introduction to human genetics cannot, at this stage of revolutionary development of our knowledge, be a compendium which presents all facts and concepts of a complete science. It can attempt merely to clarify the subject matter, the methods, and the thinking processes used in human genetics. Those who are interested in the specific results of work in human genetics are referred to the systematic presentation by Verschuer which appeared recently (1959). It is the aim of the present introduction to arouse understanding for and interest in a science which is in a lively state of flux. A person who is truly educated in science is distinguished not so much by his knowledge of material as by the ability to understand facts, judge them critically, and evaluate them in their relationships to one another.

One of the main difficulties in making a brief presentation of human genetics which can be generally understood is that examples must be borrowed predominantly from the field of medicine. Without examples, the discussion would be unclear and difficult to retain. However, it is not possible within the limits of an introduction to human genetics to give, at the same time, an explanation of the medical aspect of each example so that it can be understood by the general reader. Therefore, the book is essentially designed for the physician, who has the necessary preliminary knowledge. Nevertheless, I have tried to choose and describe the examples in such a way that a non-medical reader with some previous biological knowledge can also understand their important aspects. The appendix contains a glossary of medical terms as an aid for this purpose. The genetic discussion, on the other hand, presupposes no knowledge of genetics of any sort and begins in a very elementary manner, but not until the second chapter. In the first chapter, an attempt is made to give an over-all view of the nature and significance of human genetics. Some of the details of this portion can be understood properly only when the subsequent systematic presentation has been studied.

The basic problems of medical genetics are simple. However, we know little about the hereditary transmission of normal physical and psychic characteristics. The methods involved here are complicated, and the contributing factors difficult to survey. For the time being, opinions can be stated only with great caution. My goal would be achieved if I could convince the reader that human genetics is of extraordinary theoretical interest and of great significance for human life, not only in those areas where it is already in command of a firmly knit body of knowledge, but also in those where its pronouncements are still groping and uncertain. The reader has the right to learn which important problems have not yet been resolved and which only appear to have been solved. The procedure customarily used in some popular presentations—that of putting a protective covering over our lack of knowledge by means of a web of words—is regrettable. I hope that the admission of our lack of knowledge, which has been necessary with some of the questions raised, will cause the reader not to feel resigned but to find stimulation for new and more precise observation and for more thorough critical thinking.

The book is intended not for the specialist in human genetics but primarily for students and physicians who are interested in this field. Mathematics has been dispensed with as far as this is possible in a field where quantitative judgments concerning probabilities play an important role. I have made an effort to develop from observation the few formulas which appeared to me to be indispensable and to interpret them clearly. The chief examples of a clear presentation of the subject which have influenced the present introduction are *Menschliche Erblehre* by F. Lenz (1936), and the book which is related to it in its approach, *Introduction to Medical Genetics* (1940 and 1959) by Fraser Roberts. Whenever possible, the examples used to illustrate general principles, in addition to being suitable genetically, are also of clinical, pathological, and physiological interest. Here, special emphasis has been placed on pointing out the fruitful relationship between clinical and genetics research, as it has been expressed in a series of significant books during the past two years. The following deserve special mention: de Grouchy, *L'hérédité moléculaire, conditions normales et pathologiques;* McKusick, *Vererbbare Störungen des Bindegewebes;* Lamy, Royer, and Frézal, *Maladies héréditaires du métabolisme chez l'enfant;* and Hsia, *Inborn Errors of Metabolism.*

W. Lenz

Hamburg, 1961

CONTENTS

ILLUSTRATIONS

TABLES

GENERAL INTRODUCTION

Human genetics as a study of the causes of
individual differences

Human genetics may be defined as the study of individual differences which are based on differences in the chromosomes of the cell nucleus and in their elementary constituents, the genes. Ultimately, all manifestations of life are probably determined by the genes of the cell nucleus. However, only those manifestations of life which can be observed to be different among different individuals can be dealt with genetically. One could therefore suppose that genetics concerns itself only with a small, insignificant portion of the manifestations of life, namely, with variable and thus insignificant deviations, and that the hereditary material common to all men is not seen by the geneticist. Of course, if the human geneticist looks beyond his limited field into the study of evolution and comparative biology, he no longer observes the differences among human beings but instead those between man and his ancestors or the anthropoid apes. These differences, which lead us closer to the nature of man, are, according to all our present knowledge, also determined exclusively by differences in the chromosomes and genes. Even if the human geneticist studies only differences among people, he will soon observe not only that he is including superficial, accidental characteristics but that genetic observation, which is directed toward differences, penetrates deeply into the nature of man. This is due to the fact that all human characteristics exhibit enormous genetic differences.

These differences among people are so familiar and self-evident to us that we rarely give them any thought. Almost everyone can distinguish hundreds of persons with complete certainty. Very seldom are there two people among the many whom we encounter in the course of our lives who actually look identical. The only exception is that of identical twins, who have the same heredity and whose similarity amazes us again and again. If we ask ourselves why such a resemblance is astonishing, we realize that it is only because this is such a surprising contrast to the ordinary diversity of people. Therefore, our reaction to the sight of a pair of identical twins is one of the most notable testimonials to the unlimited variety of individuals. It is here that the scientific significance of identical twins for human genetics lies. The equality of their characteristics serves as a measure for comparison by which the meaning of hereditary differences among other individuals can be gauged. If the human race were genetically uniform, like a strain of laboratory animals which has been inbred by sibling matings through many generations, identical twins would not be conspicuous as something out of the ordinary—in fact, human genetics would be meaningless.

Pure races, such as are known by breeders of animals used in agriculture or of strains of rats used for experimental purposes, have never existed among men. Human genetics has significance to the extent that the hereditary differences among men are important. Also, because of the multiplicity of individual differences, the methods of genetics yield a coherent picture of almost all life processes. In his form and behavior, man is not a uniformly standardized type. A schematic representation of his average anatomy and his physiology, therefore, can be far removed from reality in an individual case. Thus the picture of variability which is the result of research in human genetics is a necessary supplement. Basically, our culture with its high valuation of the individual rests upon the biological multiplicity of people. The variety of individuals in the population is of obvious advantage also for the preservation of the species. Populations which include many different forms and varieties of response are better adapted to changing environmental conditions.

The definition of genetically determined differences can be limited to the differences present in the cell nucleus. We know nothing about hereditary factors in the cytoplasm of human beings. In the other mammals, also, cytoplasmic heredity is unknown. One may probably conclude from this that cytoplasmic heredity, even if it is possible in man, has no practical significance. Unfortunately, some medical books still contain the old expression "germ-plasmic." Weismann (1892) used this term to mean approximately what we now call gene-determined. According to our experience, however, this expression is misunderstood. It has no connection with the hypothesis of cytoplasmic heredity. For a time, plausible arguments appeared to exist for the assumption that the basis of mongolism could be damage of the cytoplasm of the egg cell. Meanwhile, however, studies on chromosomes have shown that mongolism also is determined by an anomaly of the nucleus, namely, by an excess chromosome.

The chromosomes—the carriers of heredity—transmit the material basis of family resemblance from the parents to the children. It is not practical, however, to base the definition of genetic properties on the resemblance between parents and children. This resemblance may be determined by heredity, but the lack of resemblance is usually also determined genetically. We include among genetically determined characteristics those which occur only with a special combination of genes. The individual genes are present in the parents, but not always in the effective combination which alone determines the characteristic. The latter occurs in the children only if the inherited genes combine in a certain way. The recessive hereditary traits are the simplest example of this kind, since they evoke the corresponding feature only if they are inherited from both parents and are therefore present in duplicate. Lack of resemblance between parents and children may also be caused by discontinuous changes in the hereditary traits, so-called mutations, which, on the whole, are not uncommon. Parents and children differ in these mutant genes, and yet these are hereditary traits which are further transmitted to the descendants exactly like the remaining traits.

Thus, the human geneticist studies the differences due to chromosomes and genes. This definition, however, is valid only for the purified end product of research in human genetics, not for its raw material. The object faced at the outset by the human geneticist consists of human differences as such, and one cannot determine at once whether they are the result of genetic or exogenous factors. Man does not live under uniform laboratory conditions in which environmental differences can be ignored. The human geneticist must at all times take into account environmental factors, which may have a disturbing influence on the results obtained with purely genetic methods.

Human genetics is an etiological science. The description and classification of individual differences are of interest to the geneticist primarily to the extent that they are a prerequisite for his actual work, the analysis of the genotype. The method of genetics is based on comparison of blood relatives of differing degree.

The further question, how the genotype produces the corresponding phenotype, must be worked on mainly by means of non-genetic methods. The so-called field of *phenogenetics*, which explores the causal relationship between trait and characteristic, is a part of developmental physiology rather than of genetics. Not only the mature individual but his entire course of development starting from the fertilized egg cell is designated as the phenotype. In the word phenogenetics, "-genetics" has a rather striking double meaning. It suggests hereditary traits and, at the same time, individual ontogenesis, primarily the morphogenesis or development of form. Phenogenetics concerns itself chiefly with analyzing the phenotype. It is thus not a part of genetics in a strict sense. Of course, it is neither possible in theory nor advisable in practice to draw sharp boundary lines between individual disciplines which treat different aspects of the same subject. Indeed, its close collaboration with biochemistry, cytology, serology, and clinical and medical research has enabled human genetics to achieve its most significant advances. As an isolated science, detached from related disciplines, human genetics would be condemned to sterility. It can best fulfil its essential task in association with other biological sciences.

The significance of human genetics for the
sciences of man

Anthropology could be regarded as a comprehensive science of man; but it cannot be clearly separated, either in its subject matter or in its methods, from the related disciplines of anatomy, human genetics, ethnology, sociology, and others. When, more than a hundred years ago, Darwin made it possible to explain the laws of the development of organisms in their basic outlines, anthropology blossomed forth under the grand illusion that it would now be possible to understand the biological and cultural development of races and cultures as a unified picture. But sober skepticism followed the premature enthusiasm over such a theory of evolution. This skepticism pointed, and rightly so, to the considerable autonomy of cultural phenomena as compared with biological facts. Strictly scientific anthropology was reduced more and more to a theory of the descent of man carried on purely morphologically, and to the study of the somatic variability of man, especially of the parts of his skeleton. The Darwinian concept of the selective effect of certain environments and of the survival of certain forms is rarely mentioned in anthropology today. The question whether certain climatic, sociological, and cultural environmental configurations breed certain hereditary traits is seldom asked, and could not be answered even with approximate exactness by the methods in present use.

For several years, it has been customary to call certain general observations about the "nature" or the "destiny of man" anthropological. Here, "man" is not the subject of scientific investigation but an image colored by ideologies or determined by certain desires. Human genetics contributes nothing to "anthropology" in this sense, aside from the admonition that pronouncements about "man" are questionable in view of the extraordinary diversity among individual human personalities. However, if anthropology is pursued not as a theory of values but as a science, human genetics is one of its most important foundations. One might think that human genetics must be the true basis of anthropology; but, at least for the time being, it cannot fill this role. Human genetics of healthy man is only in its beginning stage. Present-day human genetics must be built predominantly on the foundation of hereditary pathology. We shall become better acquainted with the reasons for this. However, as soon as human genetics has analyzed the part played by hereditary traits and by environment in the normal variability of people, it could become the true scientific basis of anthropology.

Anthropology is, on the one hand, still a descriptive and classifying science and, on the other hand, a historical one that must work with hypotheses which it is difficult to check. The laws of economic, cultural, and intellectual development are considerably better known themselves than are their biological presuppositions and consequences. But it is obvious to everyone whose eyes have once been opened to the viewpoint of genetics that, in the end, economic and cultural achievements are decisive for the fate of mankind only to the extent that they become biologically effective. If ever mankind as organized society should learn to take its fate sensibly in hand in the face of the threatening dangers of overpopulation, hunger, and atomic weapons, a knowledge of human genetics will be indispensable to the process. The human geneticist is only now beginning to decipher the alphabet in which the biology of man is written. The grammar of the language and the meaning of that which is written are still concealed from him. He can, how-

ever, imagine an anthropology of the future in which the biological assumptions and consequences of the development of cultures and races will be
understood better than is possible today. It must be admitted that we are
still far from this goal, which many, even decades ago, naïvely thought was
near.

A special branch of the biology of man, called *constitution theory,* has
been developed. This is the study of the forms of body structure and their
relationship to psychic traits and to a predisposition toward certain diseases.
Regrettably, constitution theory in its practical application is often a semi-
scientific art of divination which believes that it can rely extensively on a
subjective view of the entire man, untouched by precise genetic and cor-
relation-statistical studies. The simplifications and exaggerations, as well as
the lack of a sense of method exhibited by individual exponents of constitution theory, have contributed to its popularity but have brought it into so
much discredit in scientific circles that it has largely disappeared from medical thinking where, a few decades ago, it played a rather large role. A substantial objection to the usual constitution theory is that it does not distinguish sharply enough between the influences of hereditary traits, age, and
state of nutrition on body shape. The relationships between body shape and
predisposition to illness appear to be based primarily on relationships between nutritional state and predisposition to illness, as well as between age
and predisposition to illness. The main danger of popular constitution
theory and similar arts of divination, such as graphology and physiognomy,
lies in the fact that their practitioners believe they know something intui-
tively which must, by the nature of the matter, in many cases be exposed
as an invalid prejudice.

To be sure, even constitution theory has a valuable core. There are indeed
remarkable correlations between physical and psychic features and predisposition to illness. However, these are usually so weak that, in an individual
case, an opinion rarely turns out to be more correct when the constitutional
type is taken into consideration than when it is not. A physician who at-
taches great importance to the constitutional type often believes that he
knows at first glance how he must evaluate a patient. This places him at a
considerable disadvantage compared to another physician who reserves his
judgment. Unjustified certainty is more dangerous here than justified uncertainty. This is particularly true in regard to the popular habit of estimating character on the basis of body structure. If constitution theory is
given a more solid scientific foundation, and if it observes its limitations
more closely, it can develop into a substantial component of anthropology,
human genetics, and medicine. Valuable formulations are contained in
books by Parnell, Lindegård, and Schlegel.

Psychology is more concerned with individual differences than are anatomy and physiology. There is hardly a field of human activity in which
individual differences in talent, temperament, and character are not of great
significance. For every unprejudiced observer, the differences determined

by inherent psychic traits are especially impressive. On the other hand, powerful prejudices in effect here prevent the recognition of these differences. Practical psychology attempts to comprehend the characteristics of the individual in order to determine his capabilities and to be able to recommend a suitable place for him. It is usually assumed that psychic characteristics cannot be arbitrarily molded but that they are rather stable. In the field of psychology, the investigation of heredity is especially difficult. Physical features are more easily isolated and described quantitatively. In the psychic field, on the contrary, individual functions are integrated into complex patterns of behavior. Identification of individual psychic traits is impossible. Our knowledge of the genetic foundations of lack of intelligence and above-average ability is the farthest advanced. This also points to a problem which is especially important for society. If technical and scientific progress is to be maintained, a certain minimum of human intelligence is needed. Unfortunately, however, intelligence cannot be taught. It is well known that in numerous professions there are not enough sufficiently capable replacements. The genetic aspect of this problem should not be overlooked.

Genetic aspects of medicine

Human genetics studies the causes of individual differences. The significance of human genetics for medicine lies in the contribution it can make to research into causes, or etiology. Systematic pathology or nosology today operates as much as possible according to etiological principles. It is especially to nosology that human genetics has made decisive contributions. Genetic investigations have frequently led to the separation of a group of diseases with similar symptoms into individual, mutually independent types. The practitioner is sometimes a little annoyed at this ever continuing splitting-up of groups of diseases, which burdens his memory and which, at times, appears to be of merely academic interest. The clinician, however, and the human geneticist working in collaboration with him, can point to the fact that genetic autonomy of a disease indicates a particular pathogenesis and that the prognosis, the therapeutic possibilities, and the hereditary outlook for the family can depend decisively on the special type at hand.

The physical and psychological differences among healthy persons cannot, in general, be traced back to individual hereditary traits. They are determined by numerous traits, the joint effect of which determines individuality but no single one of which exerts an effect uniquely in contrast with the normal variety of men. Hereditary traits which do determine a clearly recognizable deviation lead to a disorder in the normal function or form, that is, they are pathological. Therefore, the basic laws of heredity can be recognized more easily from the pathological hereditary traits. It can be said that *pathological states in man are usually determined by a single hereditary trait, while normal differences, as a rule, are determined by numerous traits.* Many examples may be cited for this rule of thumb. It cannot, however, be made a general principle. The more carefully we analyze normal character-istics, and the closer we thus come to the goal of recognizing that they are composed of several primary effects of individual traits, the greater the num-ber of normal component characteristics we shall eventually be able to trace back to individual traits.

Presumably there are also diseases predominantly determined by heredity which are based on the combined effect of numerous traits. If no single example that has been quantitatively analyzed in a satisfactory way can be cited for this possibility, the reason probably is simply that the analysis of the hereditary transmission of a disease to which more than a single trait or a single pair of traits contributes presents almost insurmountable methodo-logical difficulties in the case of man. We shall return to this question later, when discussing the methods of investigation of hereditary transmission through numerous hereditary traits (see pp. 173 ff.). The only complete-ly explained cases in which two genes jointly lead to a disease are sickle cell thalassemia and a few other anomalies of the red blood corpuscles (see pp. 138 ff.). It would probably hardly have been possible to explain the mode of inheritance even for these anomalies, if it were not possible also to dem-onstrate the effects of the two participating genes individually.

One can think of the different diseases as being schematically classified according to the relative roles played in their etiology by heredity and envi-ronment. Such a concept looks approximately as follows (modified from Fraser Roberts):

1. Diseases which occur only in persons with a certain genetic constitution, and always in these, without regard to environmental conditions. (Examples: chon-drodystrophy, polycystic kidney disease, hemophilia.)
2. Diseases which occur only in persons having a certain genetic constitution, and in these not always but only when also influenced by certain environmental factors. (Example: gout.)
3. Diseases which occur in persons with differing genetic constitutions but which differ in frequency and severity according to constitution and environment. Environmental factors play an important role but become effective only in certain genotypes. (Examples: essential hypertension, duodenal ulcer.)
4. Diseases which can occur in any arbitrary genetic constitution but whose sever-ity and frequency depends on the constitution. (Examples: tuberculosis, dental caries.)
5. Diseases which can occur in every genetic constitution and whose frequency and severity depends almost exclusively on exogenous factors. (Examples: roentgen burns, lye burns.)

Usually, the etiological relationships are completely explained only in diseases of groups 1 and 5. If a disease is determined by a certain hereditary trait, only by it and always by it, then its etiology is known. This is true for all hereditary anomalies with a completely explained mode of inheritance. To be sure, one occasionally reads of such diseases that their etiology is still unknown; but this is not correct. Etiologically, a hereditary trait which has undergone a pathological mutation is responsible. What is frequently still unknown is the pathogenesis, that is, how the hereditary trait leads to the symptoms of the disease.

The purely exogenous diseases actually include only those in which the damage does not occur in the natural environment of man. As soon as one is dealing with damage with which men have always had to cope, such as hunger, coldness, infections, and solar radiation, adaptation reactions occur which are produced by selection and which often show marked individual differences. Most of the exogenous diseases therefore belong in group 4. Thus, genetic investigations can be of interest also in those illnesses which are predominantly determined by environmental influences, as in tuberculosis, dental caries, or multiple sclerosis. Even in the case of disorders which are obviously determined exclusively by environment, such as endemic cretinism, the proof of the lack of participation of hereditary factors, carried out by genetic methods, can be of importance (Eugster).

In groups 2 to 4, the interplay between hereditary traits and environment has seldom been explained adequately. Thus it is open to discussion whether gout properly belongs to group 3, that is, to those diseases which can occur in different genotypic constitutions. Essential hypertension, on the other hand, perhaps should belong to group 2. It may possibly be determined by a special hereditary trait which does not always become evident. A few extreme representatives of the psychosomatic trend even go so far as to include ulcers, essential hypertension, and schizophrenia among disorders which, in principle, can occur in every genotype.

Groups 4 and 5 are in the field of general prophylaxis and hygiene. Individual prophylaxis is possible in only a few of the diseases of group 1. In groups 2 and 3, however, individually directed prophylaxis is of decisive significance. Here the concern is to locate genetically predisposed individuals and to keep away from them those exogenous injuries which favor the manifestation of the traits. Groups 2 and 3 are of special significance not only because of the better prophylactic and therapeutic prospects which are presented by the possibility of influencing the environment but also because of their greater frequency. Individual cases of the typical purely hereditary diseases of group 1 are rare, but their total frequency is great enough so that every physician repeatedly encounters patients in this classification. The physician therefore should be familiar with the general range of problems pertaining to these purely hereditary diseases. Starting from group 1, the closer we come to group 5 the more frequent in the population is the constitution that is predisposed toward the individual groups of diseases, until, finally, in group 5 the entire population is equally predisposed.

In a discussion on human genetics, the purely hereditary diseases must occupy a disproportionately prominent place for didactic reasons—a place which can be justified more by its theoretical than by its practical significance. These hereditary diseases, however, have a pre-eminent theoretical significance, not only from the point of view of genetics, but also from the most diverse medical aspects. Probably several hundred such hereditary diseases are known, each of which is based etiologically on a single pathological hereditary trait. The individual disorders are homogeneous etiologically and pathogenically. To be sure, it has not yet been possible in every case to single out the individual, independent types among a group of similar diseases. In the end, the hereditary diseases are based on circumscribed defects of a chemical nature. The type and action of these defects will be discussed in detail below (see pp. 69 ff.). If the attempt to explain the chemical nature of the fundamental "basic defect" is successful, the nosological classification of the disease can be considered complete. But medicine has the further task of explaining the relationships between the basic defect, which should primarily be regarded as homogeneous, and the frequently quite diversified symptoms of the disease. The discovery of the pathogenesis of hereditary diseases often permits a deepened insight into the interplay of functions in the organism. The isolated, circumscribed defects may be compared to purposeful experiments of nature. Agammaglobulinemia is an impressive example of this kind. In this hereditary anomaly, gamma globulin is missing from the blood. Resistance to bacterial infections is thereby very much reduced. Studies of patients with agammaglobulinemia have contributed a wealth of important information regarding the participation of various body substances and reactions in defense against disease. Our concepts of intermediary metabolism have been considerably enriched by the observation of hereditary disorders in which individual enzymes are lacking. When the connection between basic defect and clinical symptoms has been explained, one may hope that effective means will be found with which to attack it. A truly etiological treatment is not possible with hereditary diseases. In principle, however, hereditary diseases are accessible to therapy.

In general, the *family anamnesis* is of significance for the practitioner only if another family member has, or has had, the same disease as the patient. "Hereditary burdening" with common conditions such as neuropathy, psychoses, carcinoma, susceptibility to allergies, or malformations, which have no connection with the disease of the patient, is often without clinical interest. Family anamneses should, therefore, be brought in to the extent that they are relevant to the diagnosis and evaluation of the disease. The fact that a family member has had the same disease as the patient may help in making a diagnosis earlier than would otherwise be possible. This is especially important where an early start of prophylactic or therapeutic measures can mean success in treating the disease. Thus, in the case of galactosemia, it is essential to administer a diet free of lactose, that is, without milk, starting with the first days of life, before severe disturbances of growth, liver enlargement, jaundice, and clouding of the lenses set in. In general, the diagnosis can be established at this early stage only if the disease has already been recognized in preceding siblings. In hemochromatosis, which leads to cirrhosis of the liver, diabetes mellitus, and skin pigmentations ("bronze diabetes"), this unfavorable development can be prevented by regular venesections. In the female sex this measure usually is not required, since menstrual bleeding guarantees a sufficient loss of iron, which ordinarily prevents severe manifestations. Knowledge of the family anamnesis should lead to a study of the iron metabolism, by means of which the diagnosis can be made long before the occurrence of the more serious consequences of the disease, and therefore in time for an effective prophylaxis.

For scientific purposes, family anamneses without supplementary medical examination of the family members are almost valueless. In an individual case, they may contain reliable and important pieces of information; but it is difficult to distinguish these cases from the unreliable ones. In the pages of medical literature, a number of rather fantastic pedigrees are recorded which have led to considerable confusion and complicated speculations. For example, a French family has become famous in which, for three generations, all seventy-two descendants were female. A great deal of effort was devoted to various hypothetical explanations of this "monstrous female regiment," as it was called; but it appears that no attempt was made to examine the stories of the source patient about her family by obaining objective confirmation. In another case, a pregnant woman reported to the physician that she had had nine pairs of twins by her first husband, who himself had a twin sister. The family tree was published because of its great scientific interest. Examination uncovered the untenability of the statements. Numerous textbooks on human genetics contain pedigrees with "holandric" or "hologynic" heredity. A type of heredity which affects all male descendants of an afflicted male individual, but never the female descendants, is called holandric. Hologynic heredity designates the affliction of all female, but of no male, descendants of an afflicted woman. The best-known example of holandric heredity is that of the Lambert family, which, in the eighteenth century, exhibited in four generations cases of ichthyosis hystrix gravior, an extremely disfiguring skin anomaly which led to the designation of those afflicted by it as "porcupine men." These unfortunate patients, whose skin was covered with horny bristles, were seen and described by numerous medical authorities of that period both in England and on the Continent, where they were exhibited for money. A detailed examination of the historic background of the Lambert case by Penrose and Stern (1958) revealed that this pedigree with holandric heredity was full of errors. In another pedigree with holandric inheritance of excessive hair growth on the external ear, the only evidence came from an 81-year-old patient who lived in an institution and who had previously spent three terms in an institution because of alcoholism and religious fanaticism. The evidence for hologynic inheritance, for example, of a peculiar eye anomaly (cataract?) with atypical color blindness (see Stern and Walls, 1957) or of a total anosmia in twelve women in three generations, does not look any more reliable.

Family anamneses are the more trustworthy the more they are limited to the immediate family members. Detailed data on healthy and diseased parents, siblings, and children of a patient are part of a complete anamnesis.

Usually, extensive research on the pedigree is meaningful only in the case of anomalies which are very widespread in a family. The human geneticist finds a considerable deficiency in medical literature, even in publications on conditions of obvious genetic interest, in that it contains little or no data on the nearest blood relatives. Data about healthy family members are just as important for the evaluation as are data about diseased members.

An etiological prophylaxis of hereditary diseases is possible primarily through limitation of the propagation of carriers of the hereditary trait. Avoidance of marriage between relatives, about which more will be said in the proper context, and protection from mutation-producing rays are quantitatively of less importance. In some cases, a decision not to have children must be considered necessary from the humane and social points of view. For example, the probability that the child of a patient with tuberous sclerosis will also be afflicted and thus will undergo progressive mental deterioration and suffer from severe fits is 50 per cent. No responsible physician could conceal the risk in such a case or recommend that the patient have children. Also in cases of severe recessive hereditary ailments which lead to chronic crippling or slow deterioration with the certainty of a fatal outcome, further pregnancies should be prevented. If, in a given family, a child is born with a recessive hereditary ailment, the probability of any additional child's also being afflicted amounts to 25 per cent. In general, it is not the task of the physician to force his personal opinion on the parents. Nevertheless, using all possible tact, he should tell them with complete frankness how great the risk of illness is for a possible additional child. Of course, the decision whether they want to take this risk cannot be made for the parents by anyone. The physician who knows the genetic nature of a disease is obligated, not in a legal sense, to be sure, but nevertheless in a moral one, to discuss the risk with the parents. Detailed knowledge is required for this. An impromptu genetic consultation is frequently impossible even for the specialized human geneticist without first undertaking an intensive examination of the case and consulting the original literature. Information on hereditary pathology recorded in text- and handbooks is seldom reliable enough for a responsible consultation on hereditary biology.

Frequently it is not so much a matter of issuing a warning as of dispelling doubts. In most of the severe congenital malformations, for example in mongolism, the risk of a repetition in siblings is so small that the parents can be reassured.[1] Of course, in no case may the physician say that there is no risk at all. Every pregnancy entails risk. Even if no hereditary ailments or malformations are known in the family, the chance that a child may have a severe malformation amounts to approximately 1 per cent. In general, even the layman is able to understand the definition of a risk in per cent of the probability, if one goes to the trouble of explaining it patiently. Usually, such

[1] It has recently become possible to distinguish, by means of chromosome studies, the rare cases of familial mongolism from the considerably more frequent sporadic ones (see p. 122 n.).

quantitative data have a reassuring effect because they reduce uncertain and exaggerated fears to understandable and smaller dimensions.

The core of human genetics is the statistical treatment of the frequency of features among the blood relatives of carriers of the features. One can expect a result which agrees with simple genetic hypotheses only if the feature investigated is genetically uniform, that is, homogeneous. Features or diseases which are very similar or even indistinguishable can have genetically differing causes, that is, be heterogeneous. Occasionally it is possible, by means of refined methods of investigation or more precise observation, to distinguish the different hereditary types. Often, also, a statistical genetic analysis provides the incentive for closer investigation. Maximum usefulness for research is attained when formal statistical genetics works hand in hand with the clinical analysis of features. Genetic uniformity of the material is by no means a prerequisite for its treatment by the methods of genetics. On the contrary, genetically uniform material seldom provides anything new to be explored from the point of view of genetics, whereas a methodical analysis of genetically heterogeneous material often leads to entirely new disclosures concerning the classification and nature of the disease. A fine example of this is the investigation of the three different types of progressive muscular dystrophy by Becker, or the separation of two different forms of Pfaundler-Hurler disease.

CHAPTER TWO

THE NATURE OF GENES

Transmission of genes

AUTOSOMAL GENES

The genetic individuality or the genotype of a human being is the sum of his hereditary traits, or factors. Individual factors can be transmitted independently of each other to the descendants. These units of heredity, or genes, are arranged linearly on the chromosomes. The chromosomes exhibit exactly the behavior to be expected of the material carriers of heredity: they are present in all body cells in constant, duplicate number (in the *diploid* state); in the mature germ cells, on the other hand, in single number (in the *haploid* state). The laws according to which genes are transmitted from one generation to the next are simple and generally valid. They are independent of the effect of the genes. Regardless of whether or not a gene has a recognizable effect, it is passed on in the same manner, on the average to half of the descendants. The action of genes is among the more complicated problems of the study of heredity. The transmission of genes, however, follows a simple scheme. It is therefore useful, for didactic reasons, first to explain the transmission of genes without regard to their effect. Of course, we shall gain a vivid picture of heredity only when we get to know, within the manifold possibilities of gene action, the basic rules of execution of the simple law of inheritance of genes.

Genes are present in pairs in all body cells and in the immature germ cells; each individual has inherited one of each pair from his mother and one from his father. During the formation of mature germ cells, the gene pairs again separate in a regular way, so that only one gene of each pair reaches each mature germ cell. The maturation division is also called reduction division, since it reduces the double chromosome set to the single (haploid) one. In the union of two germ cells, the haploid chromosome sets are again restored to the diploid chromosome set of the fertilized egg cell or *zygote,* from which a new individual is formed. This meaningful mechanism guarantees that the number of hereditary traits remains constant. Every germ cell contains a complete assortment of all genes; and every individual, a twofold assortment. An exception exists only in the sex chromosomes of the male, which will be discussed in more detail below. The mechanism of the transmission of genes becomes genetically significant only through the fact that the two homologous genes of a pair are not always of the same, but often of a different, nature. If the genes in homologous gene locations ("loci") of a chromosome pair are the same, we call the individual *homozygous* with respect to this gene. If the homologous genes are different, the individual is called *heterozygous*. In the latter case, the two genes are also called alleles. Thus, allelic genes are genes of which one or the other can be in a given gene locus of a chromosome. In the maturation division, allelic genes always separate. The designation of homozygosity or heterozygosity is always valid only for a specific pair of genes. A person may be homozygous in the genes for blood group A_1 (A_1A_1), but may be heterozygous in another pair of genes, for example, that for blood groups M and N (MN).

For the special case of the X chromosome, which is present only singly in the male sex, the designations homozygous and heterozygous would be inappropriate. Therefore *a gene which lies in the single X chromosome* of a male individual is called *hemizygous*. In woman, who has two X chromosomes, on the other hand, X-chromosomal genes, like all other genes, can be either homozygous or heterozygous.

Since the diploid cells in a heterozygous individual contain a 1:1 ratio of the two alleles, this ratio is preserved after the reduction division, so that germ cells with the one allele are just as frequent as those with the other. Thus, the chance for each of the alleles to undergo fertilization amounts to 50 per cent or, since it is preferable to express the number in terms of a total probability of 1 (100 per cent), 0.5. Homozygous individuals can produce only one type of germ cell in regard to the gene for which they are homozygous; 100 per cent of these germ cells therefore carry the same gene. Here, also, it is better to express the probability in terms of the number 1. The combination of genes after fertilization follows the rules of probability. The probability of coincidence of two events equals the product of the probabilities of the two single events.

If we assume that we have a coin which, when thrown, shows a picture or number, with an equal probability of 0.5 for each, then the probability of throwing the picture side twice amounts to $0.5 \times 0.5 = 0.25$, and likewise the probability of throwing the number twice is $0.5 \times 0.5 = 0.25$; the two remaining probabilities, that is, that of getting the picture on the first and the number on the second throw and, respectively, that of getting the number on the first and the picture on the second, are of equal magnitude.

Because of the diploid nature of the genotype, there are only the two single probabilities of 0.5 for each of two different alleles, or the probability of 1.0 for equal genes. The probabilities of coincidence here amount either to $0.5 \times 0.5 = 0.25$ or $0.5 \times 1.0 = 0.5$, or $1.0 \times 1.0 = 1.0$. These rules are fundamental for all of heredity. Almost all remaining laws can be derived from them. In Table 1 they are given in tabular and graphic form.

The probabilities for the individual combinations in the four sections of the table are 0.25 each. But in the first example, two pairs of sections are the same, so that, by addition, only the two probabilities of 0.5 for AA and 0.5 for AB result. In the second example, showing mating between two heterozygotes, there are three possibilities, since AB is identical with BA; the probabilities amount to 0.25 for AA, 0.5 for AB, and 0.25 for BB.

The genes lie within the chromosomes. It is therefore useful to show the formal scheme of the table more clearly by means of schematic representations of the chromosomes. Figures 1 and 2 show the heterozygous and homozygous states of the alleles in the parents, their splitting in the germ cells, and the new combination, according to the rules of probability, in the children.

As a rule, the transmission of a given gene is independent of that of another, non-allelic gene. Thus, for example, a man may have inherited from his father traits A and M, and from his mother traits B and N. But there is no connection between A and M or between B and N. The genes are in-

TABLE 1

SCHEMATIC REPRESENTATION OF THE FORMATION OF MATURE GERM CELLS AND OF THE COMBINATION OF GERM CELLS DURING FERTILIZATION

1. Mating: Heterozygous × homozygous

		Genotype AB	
	Germ cells	A	B
Genotype AA		Zygotes	
	A	AA	AB
	A	AA	AB

2. Mating: Heterozygous × heterozygous

		Genotype AB	
	Germ cells	A	B
Genotype AB		Zygotes	
	A	AA	AB
	B	BA	BB

herited, not in whole packages, but, other than in exceptional cases which will be discussed later, as individual elements. The man who has the pair of alleles AB and the other pair of alleles MN can therefore form four different germ cells, which contain the genes A and M, or A and N, or B and M, or B and N. These four possibilities are equally probable (Fig. 3). The non-allelic hereditary factors can combine freely with each other in the germ cells. This freedom, however, has a certain limitation because some groups of genes remain together more frequently than would be expected according to the rules of mere chance combination. This is known as linkage. Linked genes lie on the same chromosome and can therefore be transmitted together in the chromosome section. One can picture this somewhat like the composition of a railroad train. Railroad traffic proceeds not with independently traveling single cars but with trains in which several individual cars are combined. In the railroad train, too, the cars are not indissolubly linked to each other, but the trains can be divided and newly combined. In this process, not all cars are disconnected from each other, but, as a rule, neighboring cars remain together. The linking of genes, too, is not rigid but can be broken by an exchange of segments between two homologous chromosomes. This is called an interchange of factors or crossing over. The juxtaposition of two non-allelic genes on the same chromosome of a homologous pair is one possible consequence of the alignment of genes on the chromosomes. The other possible consequence is that each of two non-allelic genes lies on another of the two chromosomes of a homologous pair. In the second case, they show the genetic phenomenon of repulsion. They are transmitted to different descendants more frequently than would be expected on the basis of a random distribu-

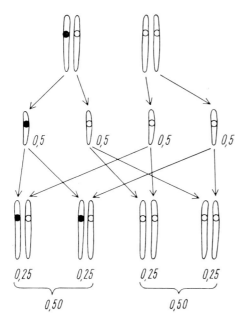

Fig. 1.—Chromosome scheme of heredity for mating between heterozygote and homozygote.

tion. In Figure 4, the state of linkage in the stricter sense (coupling), and, below, its resolution by crossing over, are represented on the left side. On the right side, linked genes in the state of repulsion, and, below, again the consequences of crossing over are shown. It is apparent from the figure that, through crossing over, genes which lie on the same chromosome become genes on different chromosomes and that, in reverse, genes on different chromosomes of a pair reach the same chromosome by means of crossing over.

The practical significance of linkage for human genetics is small. The expression "linkage," it has been found, is usually misunderstood. Linked genes do not occur together more frequently than is to be expected for chance combinations. Only if they once lie on the same chromosome in an individual do they remain together with above-average frequency in his children. On the other hand, if they lie on the two different chromosomes of a homologous pair in a given individual, they will separate with above-average frequency. Unfortunately, the nomenclature is somewhat subject to misunderstanding. The general concept of linkage applies only to the localization of genes on homologous chromosomes; the phase of coupling, that is, of placement on the

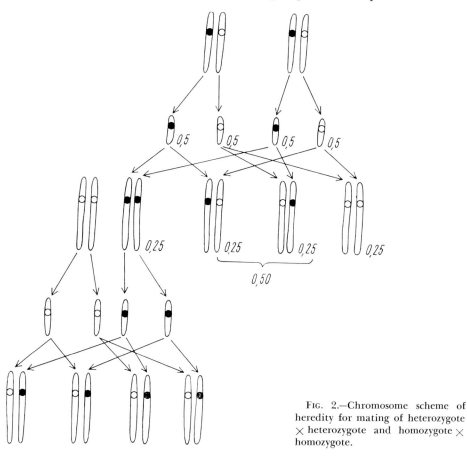

Fig. 2.—Chromosome scheme of heredity for mating of heterozygote × heterozygote and homozygote × homozygote.

same single chromosome, is distinguished from the phase of repulsion, that is, of being placed on the two partners of a homologous pair of chromosomes.

SEX CHROMOSOMES AND SEX DETERMINATION

The discussion thus far about inheritance of genes is valid for the entire set of genes in the female sex but only for 22 of the 23 chromosome pairs of the male. In the male sex, an unequal, non-homologous pair of chromosomes is present, the X and the Y chromosome, which presumably have no common

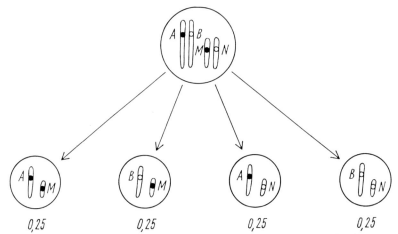

Fig. 3.—Free combination of the genes of two non-linked pairs of alleles in the germ cells

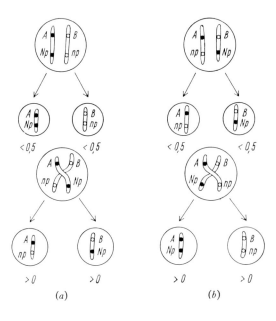

Fig. 4.—Distribution of two linked pairs of alleles to the germ cells. (*a*) Coupling phase; *below*, crossing over. (*b*) Repulsion phase; *below*, crossing over.

gene loci. The Y chromosome is responsible for the development of the male gonads. The mature germ cells of the female in every case contain 23 chromosomes under normal circumstances, namely, an X chromosome, or sex chromosome, and 22 autosomes, or non-sex chromosomes. The mature germ cells of the male also contain 22 autosomes and, in addition, either an X chromosome or a Y chromosome. Depending on whether a sperm with an X chromosome or one with a Y chromosome combines with the egg cell, a female or a male individual is produced. Since no exchange of factors takes place between the X and Y chromosomes, these chromosomes are transmitted in their entirety by the male to his children. They behave in this respect like hereditary units or genes. A child can obtain from its father either the Y chromosome, in which case it becomes a boy, or the X chromosome, which makes it a girl (Fig. 5).

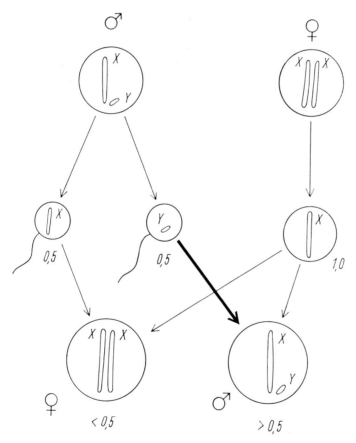

FIG. 5.—Chromosome scheme of sex determination. The heavier arrow, originating at the Y sperm, is intended to indicate that Y sperms reach fertilization more frequently, although they are formed in the same number as X sperms.

While the chance for both alleles of reaching fertilization is equally large in the case of autosomal pairs of alleles, namely, 0.5, there is a notable exception to this regular mechanism of distribution in the inheritance of sex chromosomes. For reasons which are still unknown, sperms with a Y chromosome fertilize the ovum more frequently than do sperms with an X chromosome. Therefore, immediately after fertilization the ova destined to be male clearly predominate over those destined to be female. The ova destined to be male, however, have a greater embryonic and fetal mortality rate, so that, at the moment of birth, the sex ratio is lowered to about 106 boys:100 girls. Just as we distinguish the heterozygous from the homozygous state in regard to alleles, we also distinguish for the sex chromosomes the heterogametic (XY) from the homogametic (XX) sex. In man, the male sex is heterogametic; but there are species in which the female sex is heterogametic, such as butterflies and birds.

The passage of genes through generations is altered by two factors which must still be discussed in detail—mutation and selection. Mutation means a suddenly occurring change in hereditary factors, either for which no cause is recognizable ("spontaneous mutation") or which can be caused by chemical or physical influences, especially by ionizing rays. Presumably, mutations can take place in all cells. They are of genetic significance mainly if they occur in the gametes. If a germ cell with a mutant gene is fertilized, the resulting individual has a gene not present in its parents. It can then transmit this gene to its descendants in the same manner as it transmits all other genes. The special medical significance of mutations lies in the fact that they usually reduce the viability or lead to severe anomalies. The unfavorable effect of the mutation is related to the fact that the genotype is bred by selection for adaptability. Every alteration in the complicated system which has thus been brought about has a far greater probability of leading to a disturbance of the harmony than of being useful. Mutations can affect single genes or entire chromosome segments or chromosomes. Up to now, only a few examples of anomalies in the number of chromosomes have become known in man, as well as a few cases with a chromosomal structural anomaly, namely, the attachment of one chromosome section to another, non-homologous chromosome. This is referred to as a translocation. The perfection of the techniques of cell cultures, preparation, and staining, which has brought considerable new knowledge during the last few years about the morphology of the chromosomes of man, leads one to expect that it will be possible in the future to demonstrate even finer structural anomalies. In the wild forms of many plants, and also of Drosophila, anomalies in the chromosome structure are surprisingly frequent.

The hereditary transmission of genes is affected in a decisive way by the action of the genes themselves. Genes which diminish viability or fertility are lost more or less rapidly through selection. Genes which make their carriers more viable or fertile will spread.

Chemical and biological properties of genes. Deoxyribo-
nucleic acid. The significance of ribonucleic acid.
Primary gene products

The nature of genes is a central problem in genetics, the understanding of which is important also for human genetics. However, since our concepts of the nature of genes have not been formed by observations on man, the principle of the present book must be abandoned in this instance, and we must refer back to experience which comes essentially from the genetics of bacteria and phages. Genetic properties may be transferred from one strain of bacteria to another by means of highly purified deoxyribonucleic acid (DNA). In phages, also, the ability to infect a bacteria cell and to multiply in it is characteristic only of the DNA component. In almost all living beings—bacteria, plants, and animals—the cell nucleus, in which the genetic material is

TABLE 2

DEOXYRIBONUCLEIC ACID CONTENT OF DIFFERENT
CELL NUCLEI IN mg $\times 10^{-9}$ PER NUCLEUS

	Chicken		Man
Lymphocytes	....	....	5.06
Erythrocytes	2.34	2.49	....
Liver	2.39	2.56	5.64
Kidney	2.20		6.34
Heart	2.45		....
Pancreas	2.61		5.18
Spleen	2.54		....
Sperm	1.26		2.44

contained, consists primarily of DNA. In some plant viruses, such as the tobacco mosaic virus, but also in the smaller human pathogenic types of viruses such as those of poliomyelitis and influenza, ribonucleic acid takes the place of DNA as the carrier of genetic information. Larger virus types, such as the vaccinia virus, on the other hand, usually contain DNA. The diploid cells of all organisms contain fairly constant amounts of DNA which are independent of the tissue examined but are characteristic of the species. Among mammals there are no great differences in the DNA content of the cell nuclei. This indicates that the larger body mass of man and the more complicated structure of some of his organs compared to the mouse or rat are not determined by a larger number of genes. On the other hand, the DNA content of human cells is about thirty times as great as that of Drosophila. But one should not immediately draw conclusions from this regarding the ratio of the number of genes of the two species. After the maturation division, the haploid gametes contain only about half the amount of DNA contained in the diploid somatic cells. Polyploid cells—cells with a multiple of the chromosome set—also contain a corresponding multiple of DNA.

Most geneticists today are convinced that DNA is the true carrier substance of genetic characteristics. It is often compared to a code, a secret script in which the genetic information is recorded. This secret script of nature, in which the fundamental biological characteristics from the larger virus types up to the whale are laid down, must fulfil the following assumptions:

1. It must be capable of duplicating itself identically.
2. It must contain a very high number of different pieces of information and be able to transmit them to the cells.
3. It must be able to pass by a small step from a relatively stable state into another, equally stable state (mutation).

DNA has a molecular weight of 300,000 to 5,000,000. It is composed of nucleotides located adjacent to each other in the form of chains. The nucleotides in turn consist of a sugar with 5 carbon atoms, deoxyribose, a phosphate molecule, and an organic base. In most species only four different bases

TABLE 3

MOLAR RATIOS OF PURINE AND PYRIMIDINE BASES
IN THE DIFFERENT TYPES OF DNA MOLECULES

Species	Adenine:Guanine	Thymine:Cytosine	Purine:Pyrimidine
Cattle...............	1.3	1.4	1.1
Man.................	1.6	1.8	1.0
Yeast...............	1.7	1.9	1.0
Avian tuberculosis bacterium............	0.4	0.4	1.1

occur: the two purine bases, adenine and guanine; and the two pyrimidine bases, thymine and cytosine. The individual nucleotides have a molecular weight of 300 to 350, so that about 1,000 to 10,000 nucleotides are contained in one DNA molecule. The DNA isolated from the cell nuclei of animals is not a uniform substance but a mixture of different types of molecules which, however, cannot be separated from each other because of their great similarity. In all the species that have been investigated, the DNA molecules contain two pairs of bases, present in a definite quantitative ratio: for one of the purine bases there is always a molecule of one of the pyrimidine bases, and the other purine base has associated with it the other pyrimidine base. Thus, thymine exists in a fixed 1:1 ratio to adenine, and so does guanine to cytosine. However, the quantitative ratio within the purine bases (i.e., of adenine to guanine) and within the pyrimidine bases (i.e., of cytosine to thymine) is different in different species and usually differs from 1. Within a given species, however, it is the same for all types of cells.

The structure and function of DNA have not yet been explained in detail. The most plausible model is one designed by Watson and Crick, according to which DNA is said to consist of two nucleotide chains wound around each other in a spiral; each adenine of the one chain is supposed to be joined to a thymine of the other, and likewise each cytosine to a guanine, by means of hydrogen bonds. (The Nobel Prize in medicine for 1962 was awarded to James D. Watson, Jr., of Harvard University and Francis F. C. Crick of the Cavendish Laboratory, Cambridge, England, for their work in this field.) One can picture this model somewhat like a spiral staircase in which the base pairs form the steps, and the sugar-phosphate structure the outer stair railing. This model enables the identical self-duplication, that mysterious property of genes, to be approximately understood for the first time. If the hydrogen bonds break apart and the two nucleotide chains separate, then, for spatial reasons, in a cell medium which contains the nucleotides the free bases can attach to themselves only the fitting members, namely, adenine to thymine and guanine to cytosine. By this process, the separated halves would again be able to complete the original molecule. It has already been possible to produce such self-duplication of DNA experimentally. DNA was combined with triphosphates of the different nucleosides (nucleoside = organic base + sugar). The ratio of adenine-thymine to guanine-cytosine in the

TABLE 4

COMPOSITION OF DNA IN MAN IN MOLE OF BASE
PER MOLE OF PHOSPHORUS

Base	Sperm	Thymus	Liver
Adenine..........	0.27	0.28	0.27
Guanine.........	0.17	0.19	0.18
Cytosine........	0.18	0.16	0.15
Thymine.........	0.30	0.28	0.27

newly produced DNA corresponded to the ratio in the added DNA but was independent of the ratio of the bases in the nucleosides. The DNA which had thus been synthesized in vitro without the presence of living cells was indistinguishable chemically, physically, and enzymologically from natural DNA. (In 1959, Arthur Kornberg of Stanford University School of Medicine was awarded the Nobel Prize in medicine for these investigations.) The fundamental significance of DNA for theoretical medicine now became apparent. The individuality and specificity of action of genes must evidently depend on the sequence of base pairs. Because of the size of the DNA molecules, an extraordinary number of different combinations is possible. The secret code of living organisms is written in the sequence of the nucleotides in the DNA molecules. The secret code, however, must first be translated by means of a further system into instructions which can be read by the cell. It appears that these instructions are written in the ribonucleic acid molecules.

Ribonucleic acid is distinguished from DNA by the fact that it contains ribose instead of deoxyribose, and uracil instead of thymine (thymine = 5-methyl uracil). In the three remaining bases, ribonucleic acid (RNA) and deoxyribonucleic acid agree. The base content of RNA depends on the base content of the DNA, from which it has been produced. Thus, it appears that DNA, to begin with, is able to transmit its genetic information to the related RNA molecule in a manner which has not yet been explained in detail. RNA is closely connected with protein synthesis. Even if the cell nucleus with the DNA is missing, as in reticulocytes, protein synthesis is still possible, but not in the absence of RNA. The RNA content of the cells is not constant for a given species. It is high in cells which produce a large amount of protein, and low in cells which synthesize no protein. In bacteria, one can affect the ability to synthesize protein not by means of an enzyme which destroys DNA, but rather by an enzyme which destroys RNA. In this case, the addition of RNA restores the protein synthesis. Presumably RNA supplies a kind of matrix on which the amino acids are deposited in order to be formed into specific protein substances, perhaps into enzymes, according to the specific sequence of the base pairs. (The significance of RNA for medicine also was recognized by the awarding of the Nobel Prize for 1959 to Severo Ochoa of the New York University School of Medicine.) Ochoa was able to synthesize polymeric molecules, starting with the ribose diphosphates of the four organic bases—adenine, guanine, cytosine, and uracil—which were indistinguishable from the RNA in the cell cytoplasm. They also assumed the typical double-stranded spiral shape of the natural molecules.

One could compare the DNA molecules to casting molds of which each printing shop needs only one set. The RNA molecules could then be compared to letters which are poured in the form of DNA. The specific protein substances of the cells would, in the comparison, be the letters which are printed.

The specific protein substances are called primary gene products. In general, one can assume that at least one gene is required for each specific protein. The "one-gene–one-enzyme hypothesis" perhaps represents this relationship in a somewhat too simplified form; but it supplies a satisfactory explanation for numerous metabolic anomalies in man. To be sure, probably not all primary gene products are of an enzyme nature. The polypeptides of the hemoglobin molecule, the different blood group substances, and some building blocks of connective tissue are presumably primary gene products but not enzymes. Therefore, it is preferable to speak of a "one-gene–one-specific-protein hypothesis." As we shall see, even this is not yet a simplification which is always applicable.

Human genetics can make significant contributions at present to the question of the nature of the primary gene products. For decades, the human geneticist was able merely to regard with admiration the exactness of experimental genetics, and he had to base his observations about human genetics entirely on the solid foundation of animal and plant genetics. The human geneticist was able to contribute very little to genetics in general. This situation has changed considerably in the last few years. Through the excellent investigations by Ingram and his collaborators, a fundamentally new and most important discovery of a genetic nature was made for the first time in man. This concerned the explanation of the chemical difference between sickle cell hemoglobin, other pathological hemoglobins, and normal hemoglobin. The clinical and genetic significance of these abnormal variants of normal hemoglobin will be discussed later. Here, we are concerned with the chemical nature of the difference between normal hemoglobin A (adult hemoglobin as distinguished from hemoglobin F, the fetal hemoglobin), hemoglobin S (sickle cell hemoglobin), and hemoglobin C. It was proved by analysis of the building blocks of these three hemoglobins that they differ only in a single peptide, that is, in the protein component, and in this only by a single amino acid. Peptide No. 4 of one of the polypeptide chains of hemoglobin A contains glutamic acid in a certain location. In hemoglobin S this is replaced by valine; and in hemoglobin C, by lysine.

The responsible genes are alleles. One can therefore assume that the gene for hemoglobin A may be changed by a mutation into the gene for hemoglobin S or C. If the difference in the gene product is so limited that it affects

TABLE 4A

AMINO ACID SEQUENCE OF PEPTIDE NO. 4 IN HEMOGLOBIN TYPES

Hemoglobin A	Hemoglobin S	Hemoglobin C	Hemoglobin Gs
Valine	Valine	Valine	Valine
Histidine	Histidine	Histidine	Histidine
Leucine	Leucine	Leucine	Leucine
Threonine	Threonine	Threonine	Threonine
Proline	Proline	Proline	Proline
Glutamic acid	*Valine*	*Lysine*	*Glutamic acid*
Glutamic acid	Glutamic acid	Glutamic acid	Glycine
Lysine	Lysine	Lysine	Lysine

only a single one of the approximately 300 amino acids of the globin portion of hemoglobin, it can be assumed that the responsible difference in the genes is similarly limited. Possibly, the three allelic genes differ only in a single base pair.

In the past few years, it has become possible to distinguish more than thirty different pathological hemoglobin types in man by means of electrophoresis of the hemoglobin and by chromatographic analysis of its building blocks. Undoubtedly, the better explanation of the chemical differences of these gene products and of the inheritance of the corresponding genes will, in the near future, considerably enrich and refine our concepts of the nature of gene action. Not all differences between hemoglobin types are based on the substitution of the glutamic acid of peptide No. 4 by other amino acids. In several hemoglobin types, another peptide of the same polypeptide chain is constituted differently from that in normal hemoglobin A, and in still other hemoglobin types, a third polypeptide chain is altered. The globin portion of the hemoglobin molecule consists of two identical halves, each of which is composed of two different polypeptide chains, which are designated by the letters α and β. Hemoglobin types S, C, E, and D_β are distinguished from hemoglobin A only in the β chain. But while the difference for hemoglobin S and C concerns peptide No. 4, it is located in peptide No. 26 in the case of hemoglobin E and D_β. Hemoglobin types Ho2 (Hopkins 2), D_α, and I are distinguished from normal hemoglobin A not in the β chain, but rather in the α chain. The difference between hemoglobin A, hemoglobin D_α, and hemoglobin I again concerns only a single peptide (No. 23).

Apparently, there is one allele series apiece for the synthesis of the α and of the β chain. Thus, not the entire hemoglobin molecule, but the two polypeptide chains are the primary gene products. The polypeptide chains can combine freely. Each individual who is doubly heterozygous for the two non-allelic genes Hb_α^{Ho2} and Hb_β^S can form normal hemoglobin A in addition to the pathological hemoglobins Ho2 and S, since that individual is also heterozygous for the two normal

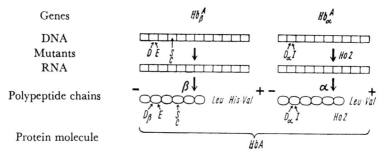

Fig. 6.—Schematic representation of genes and gene products for pathological hemoglobin types (from Allison, 1959b).

alleles Hb_α^A and Hb_β^A. Thus, he produces normal and abnormal α chains and normal and abnormal β chains. On the other hand, an individual who is heterozygous for the two alleles Hb_β^S and Hb_β^C can form no normal β chain and therefore no normal hemoglobin. In the test tube, also, the free combination of the polypeptide chains can be reproduced. If one splits pure hemoglobin C under slightly acid conditions into the molecule halves $\alpha\alpha$ and $\beta^C\beta^C$, and hemoglobin I in the same manner into the halves $\beta\beta$ and $\alpha^I\alpha^I$, and then combines the mixtures, in addition to the pathological molecules C and I and a new hybrid molecule CI, normal hemoglobin A is regenerated. Figure 6 is a schematic representation of the concepts which have just been outlined of genes and primary gene products for the hemoglobin types.

CHAPTER THREE

SIMPLE EFFECTS OF GENES

Relationships between gene and characteristic.
Genotype and phenotype

Human genes are not accessible to direct observation. We merely draw conclusions about the presence of certain genes from the presence of certain characteristics in an individual or in his relatives. In individual human beings, we first detect the recognizable characteristics whose sum constitutes the phenotype. Under favorable conditions, we can draw conclusions regarding the genotype from our knowledge of the laws of heredity or from observations of family members. Apart from anomalies in the number of chromosomes, however, the genotype itself cannot be directly determined in man.

In the literature about dermatoglyphics of the fingers, one sometimes encounters the misleading designation of genotype for a certain formula for characterizing the phenotype. The genotype which forms the basis of these dermatoglyphics is not even known. It is also wrong to equate external appearance and phenotype. Thus, for example, one may read that the phenotype of certain intersexes is entirely female, but that they have internal testicles. The meaning is, of course, that they are female in their external features. Obviously, the internal testicles are no less a part of the phenotype because they are concealed.

An understanding of the formal inheritance of genes as we have discussed it is basic for any understanding of genetics. But the formal scheme will appear vividly before us only if we follow the way in which it governs the visible effects of genes.

Interaction between alleles.—The effect of most of the genes which have been thoroughly studied genetically depends, besides on their own composition, primarily on the composition of the allelic gene, and very much less on the remaining genotype. This statement, however, cannot be made a general principle of gene action; instead, its validity is based largely on the fact that it is very difficult to find a genetic explanation for gene effects which are dependent on numerous other genes. At any rate, the interaction between alleles is of decisive significance for hereditary pathology. The fundamental concepts of dominance and recessivity are valid only for the relation between the effects of two allelic genes. Dominance and recessivity designate not special modes of transmission of the genes but special modes of their effects which make it possible to recognize the mode of inheritance.

Dominance. Effect of a gene in the heterozygous state

In accurate usage, a gene is called dominant if it exerts a recognizable effect even in the heterozygous state and if this effect cannot be distinguished from that in the homozygous state. In this sense, blood group A_1 is dominant over group O. A person of genotype A_1O cannot be distinguished by serological methods from a person of genotype A_1A_1. The distinction is possible under certain circumstances, if the nearest blood relatives can be examined. If someone has genotype A_1A_1, trait A_1 must also be present and recognizable in both parents as well as in each child. Thus, if an individual of phenotype A_1 has a parent or a child of phenotype O or B, it follows that his genotype must be A_1O (see Fig. 7).

Recessivity is a concept complementary to that of dominance. If a gene exerts the same effect in the heterozygous as in the homozygous state, it follows that the other allele is not recognizable in the heterozygous state; it is

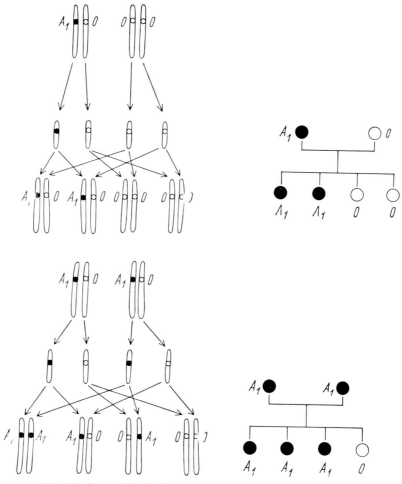

Fig. 7.—Inheritance of a dominant gene. *Left*, genotype. *Right*, phenotype

masked by the dominant gene. We call it recessive. Thus, a recessive gene is recognizable only in the homozygous state, at least to the extent that the concepts are handled rigorously. In the sense indicated, dominant or recessive genes are rare. *In medicine, the designation dominant or recessive is usually not applied in the strict sense. Here, those anomalies are also called dominant which are determined by an abnormal gene in the heterozygous state, without taking into consideration whether the heterozygous is really equal to the homozygous state.* Most dominant anomalies are so rare that the homozygous state is still unknown. In order that a dominant gene may occur in the homozygous state, two persons, both of whom possess the gene, would have to marry. But among the children of such a marriage only one-fourth would be homozygous. In the great majority of cases, dominant anomalies are inherited only from one parent. These cases show characteristic pedigrees from which the simple facts of the inheritance can be recognized directly. In regular dominance, gene and phene correspond; the genotype can be read from the visible characteristic. Pedigrees with a dominant hereditary disease therefore present direct pictures of the path which a single gene takes through the generations.

In the following figures, some examples of dominant anomalies are shown. The dominant hereditary interphalangeal synostosis of Figures 8 and 9 has become famous in human genetics because Drinkwater was able to show its

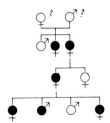

Fig. 8.—Dominant hereditary phalangeal synostosis. In this type, the proximal interphalangeal joints have undergone synostosis. In another dominant type, only the distal interphalangeal joints have undergone synostosis. (Own observation.)

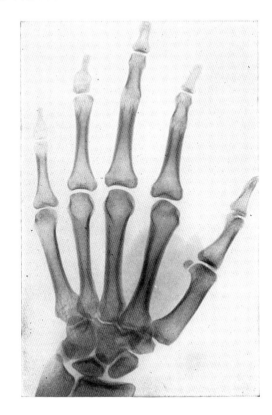

Fig. 9.—Interphalangeal synostosis. Member of the kindred of Figure 8.

probable inheritance through thirteen generations. John Talbot, the first
Count of Shrewsbury, who was born in 1390 and who plays a prominent role
in Shakespeare's drama *King Henry VI*, had this anomaly. He was killed in
the battle of Chastillon near Bordeaux in 1453. The chronicle describes
exactly the manner of his death: his thigh was shattered by a cannon ball,
and he was killed by a blow on the skull with a battle-ax. His body was sent
to England and buried in the family vault. In 1874, the grave was restored.
At that time, a friend of the family, to whom the occurrence of the finger
anomaly in the last three generations was known, examined the skeleton. He

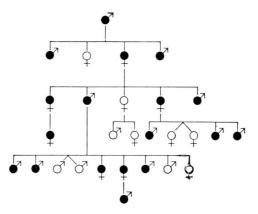

Fig. 10.—Dominant inheritance of postaxial polydactyly (doubling of the little
finger and of the little toes) in a Norwegian kindred. Excerpt from a larger
pedigree with 37 afflicted members. (From
Sverdrup.)

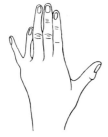 

Fig. 11.—Hand and X-ray
picture of a child's foot, from
members of the kindred of
Figure 10.

found confirmation of the data in the chronicle: the right thigh was frac-
tured, the occiput showed an extended injury from a blow, and the proximal
phalanges of the hands showed the same fusion that was present in the
descendants of the twelfth, thirteenth, and fourteenth generations. Judging
from the X-ray pictures, the anomaly was of the same type as that in the
family represented by Figure 8. The conformity of the anomaly in all
afflicted family members is characteristic for such regularly dominant anomalies. Small deviations within this common pattern do occur; thus, only
some members of the family of Figure 8 also had stiff elbows.

In the family represented by Figures 10 and 11, all afflicted members had
polydactyly. There were some individual differences, however. Thus, six instead of five metatarsals were sometimes present; in other cases, the fifth

metatarsal was split in a Y-shape; finally, some persons had only five metatarsals, but instead a Y-shaped split in the proximal phalanx of the fifth toe. The pathological features in the hands also varied. Either the fifth fingers were regularly doubled, or there was only a small appendage on the ulnar edge of the hand. In certain individuals, finally, all indications of doubling of the fifth finger were entirely absent. It is characteristic, however, that, in principle, the gene always leads to the same deviation. Thus, all the thirty-seven afflicted members had only a postaxial polydactyly—a doubling tendency on the ulnar side of the hand—but never double thumbs. Conversely, in a Swedish kindred with forty-one afflicted members, in each of five generations preaxial polydactyly—a tendency toward doubling and tripling of the thumb —was observed, but no family member had any doubling of the little finger (Nylander).

The third pedigree, also, which shows the regularly dominant form of

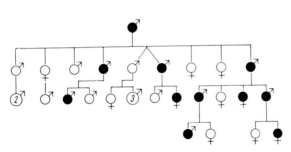

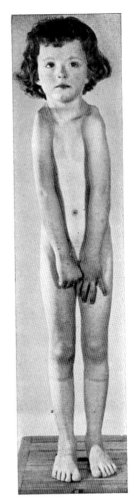

Fig. 12.—Family with cleft hands and cleft feet. (From Liebenam.) In addition to the type with regular dominance, there are others, which differ from this genetically as well as morphologically.

Fig. 13.—Dysostosis cleidocranialis. As a result of absence of collar bones, the shoulders can be brought together in front.

cleft hand, shows very similar malformations in all afflicted members in spite of a certain individual variability of expression. Such examples, which illustrate the specificity of action of individual genes, could easily be multiplied. There is a large number of different forms of brachydactyly, syndactyly, and more complicated malformations; these are based on specific dominant genes which, in principle, always express themselves in the same way. For the evaluation of the family anamnesis, the result of the specificity of the gene effect is that, if an anomaly occurring among the nearest relations is fundamentally different morphologically from the anomaly of the patient, it generally is of no significance.

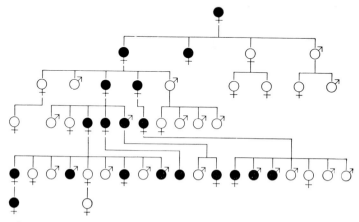

FIG. 14.—Excerpt from a pedigree of a family with dysostosis cleidocranialis. Only healthy descendants of healthy family members are omitted. (From Kahler.)

The dominant mode of inheritance can be recognized by the following features:

1. The dominant characteristic is also present in at least one parent of the carrier of a trait (exception: mutations).
2. The frequency of the trait among the parents is equal to that among the children and equal to that among the siblings (50 per cent).
3. In the case of regularly dominant traits, the descendants of unafflicted family members are likewise free of them. They do not "carry a hereditary burden" because of the presence of the trait in parents or siblings.

The test of whether a trait really occurs in 50 per cent of the parents, siblings, or children cannot be undertaken simply by using material which has been collected without special precautions. If carriers of a trait are detected through the afflicted children, one cannot simply determine the percentage of those afflicted among the children of these carriers. Only the carriers who have at least one afflicted child would be detected, but not those who happen to have only healthy children. By this method, in families with one child, 100 per cent of the children would be afflicted; in families with two children, 66.7 per cent; and in families with three children, still 57.2 per cent instead of the theoretical 50 per cent. This can be represented by a simple scheme (see Table 5).

The same considerations are valid, of course, for the frequency of the trait among the siblings of probands. In human gentics, a proband is a person who, on the basis of a certain trait or a certain anomaly, is used as the starting point for genetic investigations. Sibships with probands are

TABLE 5

RATIO OF AFFLICTED TO NON-AFFLICTED INDIVIDUALS AMONG
THE CHILDREN OF CARRIERS WITH A DOMINANT TRAIT

THEORETICAL	INCLUDED IN SELECTION STARTING WITH AFFLICTED CHILDREN			
	Uncorrected		Corrected	
	Afflicted	Free	Afflicted	Free
Families with Two Children				
●● ●○ } Included	2	0	2×1=2	0
	1	1	0	1
○●	1	1	0	1
○○ Not included	...	...	....	...
Total	4	2	2	2
Families with Three Children				
●●●	3	0	3×2=6	0
●●○	2	1	2×1=2	2×1=2
●○●	2	1	2×1=2	2×1=2
○●● } Included	2	1	2×1=2	2×1=2
●○○	1	2	0	2
○●○	1	2	0	2
○○●	1	2	0	2
○○○ Not included	...	...	....	....
Total	12	9	12	12

● = afflicted;   ○ = free.

selected in such a way that they contain at least one proband. They are therefore not characteristic of all sets of siblings which could contain the anomaly in question because of the genotype of the parents, but they are enriched by carriers of the trait through selection. This artificial enrichment can, however, be corrected by a simple arithmetic procedure. It is merely necessary to count only the siblings of the probands, but not the probands themselves. The presence of the anomaly in the siblings, of course, is independent of the question whether or not the proband is afflicted, but depends only on the possession of the responsible gene by the parents. Thus, the proband serves only as an indicator for families with the corresponding parental genotype.

The way in which the sets of siblings have been collected must be noted in the correction procedures for rectifying the artificial enrichment of the sibships with carriers of the trait. If they have been detected starting with afflicted parents, no correction is needed. In that case, those sets of siblings are included which happen not to include any carriers of the trait. If all carriers in a population are included, one can simply count each carrier once as proband in the sibships with carriers of the feature and find by addition how many diseased siblings he has. In a family with three diseased children, then, starting with each of the three probands thus counted, one would count two, or a total of six afflicted siblings. Correspondingly, in a family with two afflicted and one healthy sibling, for each of the two probands one afflicted and one healthy sibling should be counted, i.e., a total of two afflicted and two healthy ones, etc. This method of counting, however, is permissible only if all sets of siblings in the population which contain at least one carrier of the trait have actually been included. For, if only a part of the sibships has been included, the probability of inclusion for a single set of siblings is approximately proportional to the number of afflicted members. In this case, one could use the same method of counting for each carrier included as proband, but not for the siblings included by way of the probands. The more precise reason for this will follow in the discussion of the recessive mode of inheritance. The dominant mode of inheritance usually can be recognized rather simply without complicated statistical procedures. In the recessive mode of inheritance, on the other hand, flawless handling of the statistical methods is important.

There are various reasons why a dominant trait shows deviations from the frequency of 50 per cent in parents, siblings, and children of afflicted persons, deviations which are not due to errors in inclusion that can be corrected but which have other causes. Among these, irregular dominance should be named first. A gene is called irregularly dominant if, in the heterozygous state, it sometimes exerts a recognizable effect but sometimes does not. We shall be concerned in more detail further on with irregular dominance. If the frequency of a dominant gene in the parents of those afflicted remains below the expected 50 per cent—that is, if one of the two parents does not always show the feature, but the same feature occurs in 50 per cent of the children of those afflicted—then one must consider the possibility of new mutations. Indeed, a considerable percentage of all severe dominant anomalies is due to new mutations. In these cases, of course, one will find the expected 50 per cent neither in the parents nor in the siblings, but in the children. Mutations of this kind can also be simulated by extramarital conception. The human geneticist cannot simply assume the identity of the legitimate father with the biological one. If one realizes that approximately 10 per cent of all children in Germany are born out of wedlock, and that approximately 50 per cent of the firstborn children of married parents were conceived before marriage, one will count on the possibility that the officially recorded pedigree does not always reflect the true path of the genes.

The examples of dominant heredity cited so far and the subsequent quantitative considerations are valid for the case of regular dominance. In regular dominance, the gene in the heterozygous state always leads to the trait. For the anomalies named, however, the homozygous state is unknown, so that we cannot ascertain whether it is the same as the heterozygous state. In such a case, we speak of conditional dominance. If, on the other hand, the heterozygous and homozygous states are indistinguishable, we speak of complete dominance. The expression "complete dominance" thus refers not to the regularity of manifestation of the gene, but to the phenic equality of the heterozygous and homozygous states. The homozygous state is rare. It can occur only if both parents have the gene in question. This hardly ever hap-

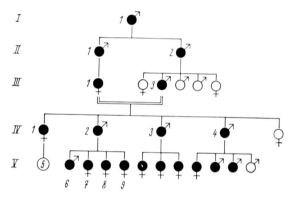

Fɪɢ. 15.—Pedigree with dominant hereditary congenital cataract. (From Komai.)

pens with rare genes. If it does occur, and if the homozygous state is indeed indistinguishable phenotypically from the heterozygous, then it is usually impossible to decide whether an individual is really homozygous or heterozygous. The genotype of these homozygous individuals can be recognized only in their children, and even there only with a certain probability. A homozygous individual passes on one of the two alleles to each child. Thus, all children must exhibit the trait. But one may not conclude in reverse that a person is homozygous if all children show the trait. Even if someone has four children all of whom show the same dominant trait, such a distribution is not too rare even for heterozygotes: 6.25 per cent of all families of heterozygotes with four children would contain only afflicted children. This number results from the multiplication of the individual probabilities for the four children $(0.5 \times 0.5 \times 0.5 \times 0.5 = 0.0625)$. One of the rare kindreds which demonstrate with considerable certainty the equality of the heterozygous and homozygous states for a pathological gene is reproduced in the pedigree of Figure 15. In this family, a congenital cataract was transmitted through five generations to nineteen members. In the third generation of the pedigree, an afflicted man married his equally afflicted cousin. Of the five children of these two carriers of the gene, four were likewise afflicted. Two of these were definitely heterozygous, since they had some healthy children (IV$_1$ and IV$_4$), but the two others were presumably homozygous, since they had only children with cataracts. It is true that this assumption would be more certain if the number of children having the hereditary illness exclusively were larger. In the family described, no difference in the extent of clouding of the lens was found between the presumably homozygous and the heterozygous carriers of the gene.

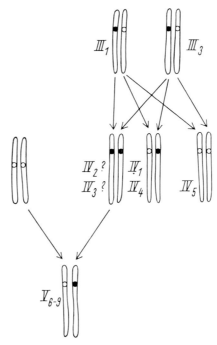

Fig. 16.—Chromosome scheme for the pedigree of Figure 15. Individuals IV$_2$ and IV$_3$ are presumably homozygous.

The mere fact that the afflicted children of two parents afflicted with a dominant hereditary disease always show the same degree of expression of the pathological feature is not sufficient for drawing the conclusion that complete dominance exists. It would also be possible that the fertilized ova with the homozygous gene are so seriously affected that they do not develop. If a gene in the homozygous state leads to death at an early stage of embryonic development, of course, only two genotypes are to be expected among the full-term children: the heterozygous and those which are homozygous for the normal allele. This could be the case in chondrodystrophy; but it can also be completely dominant. Marriages between two chondrodystrophic dwarfs have occurred numerous times without the occurrence of an especially severe form of chondrodystrophy in the children at any time. In principle, one could distinguish between the complete dominance of the gene and the homozygous lethal effect, even without knowledge of the descendants of the questionable homozygotes, namely, by studying the ratio between normal and diseased children from marriages between two heterozygotes. In the homozygous lethal effect, one healthy child would be expected for two diseased ones; in complete dominance, there would be three diseased children for one healthy child. In chondrodystrophy, only scattered observations on marriages between heterozygotes are available, so that a decision by this method is not possible.

Most dominant anomalies in man are conditionally dominant. This is no final classification. Presumably, it will become apparent as soon as the homozygous state becomes known, that incomplete or intermediate dominance is the rule. For intermediate dominance, we have a series of impressive examples from human hereditary pathology. Extremely rare cases are involved here, since the chance is small that two carriers of a rare dominant gene should happen to marry; nevertheless, the theoretical significance of these cases justifies a more detailed description. This is useful also for didactic reasons, since the conditions of recessive hereditary transmission become simpler if the nature of intermediately dominant gene action is first explained (see Table 6).

There are two different types of cystinuria-lysinuria, neither of which has any connection with cystinosis. The second, more frequent, type is recessive. Clinically, a distinction is hardly possible. The dominance of the first type of cystine-lysinuria is irregular.

In some cases on which Table 6 is based, the homozygous children came from marriages between relatives, as in the families with arachnodactyly, brachymesophalangy of the index finger, chondrohypoplasia, Pelger-Huët's leukocyte anomaly, and Willebrandt-Jürgens' thrombopathy. This is characteristic inasmuch as the chance that a gene becomes homozygous is considerably increased by consanguineous marriages. This is also the basis of the significance of marriages between relatives for recessive hereditary diseases (see pp. 55 ff.). In general, the risk that a child with a hereditary disease will be conceived is not very great in marriages between cousins, even though somewhat greater than in a random marriage. But if two candidates for marriage who are blood relatives have the same physical anomaly, the marriage should be advised against, even though the anomaly may appear to be quite insignificant. In the homozygous state, the gene can evoke considerably more severe consequences.

We have seen that dominance and recessivity are complementary concepts. If a gene is dominant, its allele is recessive. This is first of all true for complete dominance (heterozygous and homozygous states of the gene are equal). What is the situation regarding the recessivity of the normal allele in the case of hereditary diseases with intermediate dominance? If a gene with intermediate dominance causes milder pathological symptoms in the

TABLE 6

INTERMEDIATE DOMINANCE IN MAN

Anomaly	Heterozygous State	Homozygous State*
Arachnodactyly (Marfan syndrome)	Long, thin limbs, hernias, kyphoskolioses, aneurysm of the aorta, lens luxation	Especially severe expression?
Brachymesophalangy of the index finger	Shortening of the second phalanx of the index finger	Absence of fingers and toes; death at 1 year
Chondrohypoplasia†	Diminished body height by about 10 cm, short limbs, without pathological significance	Extremely short limbs, ulnar deviations of the hands, knees not completely extendable
Cystinuria-lysinuria (one form)	Low, irregular excretion of cystine, lysine, arginine, and ornithine in the urine. Occasionally cystine stones in in the kidneys	Greater excretion of the four amino acids; cystine stones in the kidneys
Ehlers-Danlos syndrome	Excessive elasticity of skin and hyperextensible limbs; wounds heal with difficulty; scar formation on knees and elbows, hematomas; tendency toward luxations	Especially severe expression of the syndrome
Elliptocytosis	Red blood corpuscles of elliptical shape, usually without pathological significance	Severe hemolytic anemia; improvement after excision of spleen
Hemochromatosis	Increased iron resorption, raised serum iron level; melanoderma, hepatosplenomegaly, diabetes mellitus, hypogenitalism, cardiac insufficiency	Earlier onset, especially severe course, cardiac insufficiency is most prominent
Hypercholesterolemia	Raised serum cholesterol; tendency toward atheromatosis and xanthomas	Extreme hypercholesterolemia; early occurrence of extensive xanthomas
Osler's telangiectasia haemorrhagica	Multiple telangiectases; nose bleeding, arteriovenous aneurysms in the lungs	Widespread telangiectases and hemangio-endotheliomas of skin, mucous membranes, lungs, spleen, liver, intestines, kidneys, and brain; death at 2.5 months
Pelger-Huët's nuclear anomaly of the leukocytes	Disorder in maturation of segmented nucleated leukocytes, which are plump with rodlike nuclei, ranging to two-segmented, shaped like eyeglasses; without pathological significance	Only neutrophilic leukocytes with round nuclei, with abnormally coarse nuclear structure; pathological significance questionable
Sebaceous cysts of the skin	Numerous sebaceous cysts of the skin	Sebaceous cysts, death in infancy
Thalassemia (Cooley's, Mediterranean anemia)	Thalassemia minor; microcytic anemia, polycythemia, target shape of the red blood corpuscles, ovalocytes, 10–20% fetal hemoglobin, increased hemoglobin A_2; occasional mild splenomegaly	Thalassemia major; hypochromic anemia which sets in early, is severe, often leads to death in childhood; pronounced spleen enlargement, hyperplasia of bone marrow, roentgenologically extended spongiosa; normoblasts in peripheral blood, poikilocytosis, anisocytosis, target cells; 40–80% hemoglobin F
Willebrandt-Jürgens' thrombopathy	Disorder in release of thrombocyte factor 3; tendency toward bleeding and hematomas; life expectancy not noticeably diminished	In early childhood, lethal intestinal bleeding or bleeding after slight injuries

* The homozygous state is clinically more severe than the heterozygous state.

† The phenotype for the homozygous state of the gene of chondrohypoplasia is strongly reminiscent of dominant chondrodystrophy; however, the head is not involved as in genuine chondrodystrophy.

heterozygous state and more severe ones in the homozygous state, it follows that the normal allele in the heterozygous state is not completely recessive but that, at least, it guarantees a condition which is closer to the norm.

In the case of hereditary diseases with intermediate dominance, a gene-dose effect has sometimes been discussed in which it was supposed that a gene causes slight damage in a single dose, that is, in the heterozygous state, but more severe damage in a double dose. This is a misapplication of the concept of dose effect. The responsibility for the difference between the heterozygous and homozygous states presumably does not lie in the single or double dose of the pathogenic gene but rather in the presence or absence of the normal allele. At any rate, this latter concept corresponds more closely to our general ideas about the effect of genes. The question whether a dose effect exists in the strict sense can ordinarily be decided only for X-chromosomal genes, the normal allele of which is absent in the male sex. If a pathological gene exerts a different effect in the hemizygous than in the homozygous state in women, one could speak of a dose effect. But any such observations on human beings are unknown.

A recessive gene which makes itself known beside its dominant allele is, of course, no longer completely recessive. One could call it incompletely recessive, or one could also, since it is recognizable when heterozygous, speak of incomplete or intermediate dominance. The concepts of dominance and recessivity start to become relative here and to lose their original meaning. Nevertheless, one can attempt to agree on a nomenclature which can be used in practice. In medicine it is appropriate to start with the pathological condition, or occasionally with a rare normal trait which is of special medical interest. The following definitions appear most appropriate for clinical use:

Dominance: the gene is of clinical significance in the heterozygous state. The case may be:

Complete dominance: the heterozygous cannot be distinguished from the homozygous state;

Conditional dominance: the homozygous state is still unknown; or

Intermediate dominance (semidominance, or partial or incomplete dominance): the homozygous is more severely pathological than the heterozygous state.

Up to now, we have discussed only those genes which always manifest themselves in the heterozygous state. If a gene manifests itself in the heterozygous state usually, but not always, we speak of *irregular dominance.* Cystine-lysinuria and hemochromatosis exhibit irregular dominance in this sense. Precise chemical investigation, to be sure, shows the dominance to be regular here also. For didactic reasons, irregular dominance will be discussed later, in connection with the additional factors which can influence the manifestation of a gene. Unfortunately, the nomenclature has been somewhat confused by the fact that Goldschmidt uses the expression "conditioned dominance," deviating from the ordinary usage which goes back to Levit and which is employed here. Goldschmidt defines "conditioned dominance" as the dominance of a gene which is influenced by other genes or environmental factors, that is, what we understand here by "irregular dominance." In common English usage, that type in which the homozygous state is unknown is designated as "conditional dominance."

I would like to reserve the concept of intermediate dominance for conditions in which the heterozygous state can be interpreted at least sometimes as a quantitative intermediate state between the homozygously normal and the homozygously pathological states. In addition, there are two further possible grades of dominance which should rather be described as qualitative peculiarities of the three states:

Codominance (combinance): the heterozygous state depends on the coexistence of two different gene products. The features of the two homozygous states can be shown side by side; one could not speak of an intermediate position. F. Lenz (1938) has suggested the expression "combinance" for this. Since Hadorn defines combinance as the functional interaction of the effects of non-allelic genes, the expression is ambiguous. It is avoided here.

Interference: compared to the two homozygous states, the heterozygous state exhibits peculiarities. These can be explained neither by establishing a quantitative average nor by simple coexistence of the two gene products; instead, through the combined effect of the two alleles, something new has been created which cannot immediately be derived from the effects of the individual genes.

Blood groups provide the best example of codominance. Antigenic properties of the red blood corpuscles are defined as blood groups or blood factors. These can be demonstrated by specific antibodies and, in part, can also be demonstrated to exist in other body cells and in secretions. In general, specific antibodies contained in the serum cause agglutination only of blood corpuscles having a specific gene-caused antigenic property; but here, also, there is some overlapping. There are antisera which react simultaneously with several antigenic properties, but apparently only with those which are determined by allelic genes. The antigenic properties of the red blood corpuscles are determined by individual genes which express themselves even in the heterozygous state. In this process, alleles express themselves in a codominant way. Thus, in heterozygous individuals, the antigenic properties A and B, M and N, K and k (Kell and Cellano), Fy^a and Fy^b (Duffy), each of which is determined by one pair of alleles, can be demonstrated to exist side by side. As long as only one antiserum is available which is directed against only one of the two allelic blood group properties, this property can be designated as simply dominant. Thus, for example, from 1946 to 1949, the Kell factor could be considered as completely dominant. During these years, only one serum was available which reacted with genotypes KK and Kk, but not with kk. Later, a new antiserum was found which reacted with an antigen "Cellano" that had been unknown up to that time. By means of systematic studies of families with K-positive and K-negative individuals, it was found that all K-negative individuals reacted positively to the new serum, while the K-positive individuals had partly positive, partly negative reactions. K-positive individuals who had K-negative children, however, always had a positive reaction to the new serum also. Thus it was proved that one was dealing with a serum directed against k, which originally had been regarded as recessive. Owing to the refinement of the methods of investigation, factor k turned from a recessive into a codominant one.

Alleles of the sickle cell series exemplify codominant pathological genes. The genes which are responsible for the production of normal adult hemoglobin, and of hemoglobin C, belong to a series of alleles, Hb_β^A, Hb_β^S, and Hb_β^C.

TABLE 7

CODOMINANT GENES. EXAMPLES FROM
BLOOD GROUP GENETICS*

Genotype	Phenotype of the Red Blood Corpuscles	Red Blood Corpuscles React with:
AA	A	anti-A
AB	AB	anti-A; anti-B
BB	B	anti-B
MM	M	anti-M
MN	MN	anti-M; anti-N
NN	N	anti-N
KK	K ("Kell")	anti-K
Kk	Kk	anti-K; anti-k
kk	k ("Cellano")	anti-k

* In all three blood group systems, there are additional alleles besides those cited, which were omitted for the sake of clarity.

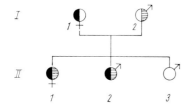

⬒ = hemoglobin C trait

◑ = sickle cell trait

FIG. 17.—Family with sickle cell hemoglobin-C disease. (From Itano and Neel.)

The corresponding hemoglobin types Hb A, Hb S, and Hb C can easily be separated from each other by paper electrophoresis. Since each individual has only two alleles from each series of alleles, the following hemoglobin phenotypes are possible: A; A + S; A + C; S; S + C; and C. The following survey provides information about the most important features of the homozygous and heterozygous states of the Hb_β series of alleles.

From the genetic point of view it is notable that codominance in the blood group characteristics and in the hemoglobinopathies presumably is of a different nature. While the two gene products for the hemoglobinopathies are different molecules which can be separated from each other, the codominant blood group characteristics appear to be present side by side in the same molecules. It is true that this cannot be proved for the blood group characteristics which cannot be separated from the blood corpuscles, and which can be identified only on the cell surface. In addition, however, there are blood group characteristics which, under certain conditions, also occur in the saliva and in other body fluids. The ABO properties are among these. It has been possible to prove that, in individuals of genotype AB, the macromolecules found in the body fluids have the antigenic properties A and B at the same time. One can make an assumption here, for which there are additional grounds (see p. 136), that the blood group substance is not the primary gene product of genes A and B, but that these genes impress their mark, so to speak, only on the gene product of another, non-allelic gene.

Interference in the sense of the above definition of a combined effect of two alleles, whose nature cannot be deduced from the effects of the two indi-

TABLE 8

CODOMINANT GENES OF THE Hb_β SERIES OF ALLELES

Genotype	Phenotype of Hemoglobin	Clinical Phenotype
Hb^A_β Hb^A_β	100% normal Hb A	Normal condition
Hb^A_β Hb^S_β	60% Hb A + 40% Hb S	Red blood corpuscles take on sickle shape under reduced oxygen tension; tendency toward hematuria; spleen infarcts during high-altitude flights
Hb^S_β Hb^S_β	100% Hb S and alkali-resistant fetal hemoglobin	Moderate anemia, scleral icterus, no tumor of spleen; attacks of weakness, abdominal pains, pallor, icterus, liver and spleen enlargement, especially after infections; stunted growth, asthenic habitus, aseptic bone necroses, abscesses of the tibia
Hb^A_β Hb^C_β	25–40% Hb C, the remainder Hb A	Slight hypochromia of the red blood corpuscles; 4–30% target cells
Hb^C_β Hb^C_β	Hb C, low percentage of fetal Hb, no Hb A	Anemia, splenomegaly, arthralgias, attacks of abdominal pain, 50–100% target cells
Hb^S_β Hb^C_β	50–67% Hb C, the remainder Hb S	Less severe hemolytic anemia than in sickle cell homozygosity, distinct splenomegaly; attacks of abdominal pain, hematuria; life-threatening crises especially during pregnancy

vidual genes, has been determined in man so far only in the haptoglobin types of the serum. In the serum haptoglobin types, one is dealing with differences in the haptoglobins belonging to the alpha$_2$ fraction of globulin; these haptoglobins are characterized by the fact that they combine with hemoglobin. In type 1, there is only a single haptoglobin fraction with fast electrophoretic migration. Type 2, on the other hand, exhibits an entire series of bands all of which migrate more slowly. Type 1 corresponds to the homozygous state of a gene Hp^1, and type 2 to the homozygous state of the allele Hp^2. The heterozygous state does not simply lead to an addition of the fast bands of type 1 and the slower bands of type 2, but to an altered electrophoresis spectrum in which the fast-moving component of type 1 is recognizable, but added to which are several components with altered speed of motion relative to type 2. Allison (1959a) recently developed a very plausible hypothesis for the interpretation of this peculiar phenomenon. According to this hypothesis, gene Hp^1 produces a homogeneous type of molecule which cannot combine with like molecules to form higher aggregates, but can do so with the gene product of gene Hp^2, but always in such a way that only one molecule of the first kind can enter into a molecular bond. The gene product of Hp^2, on the other hand, is so constituted that the individual molecules show a tendency to join together into polymers, that is, larger molecular bonds (dimers, trimers, etc.). This hypothesis makes understandable the different fractions which can occur under the influence of a single homozygous gene and the modification of these fractions by the addition of a different kind of gene product (see Figs. 18 and 19). Investigations with the ultracentrifuge have also shown that the haptoglobin of genotype Hp^{2}Hp^1 cannot be interpreted simply as a mixture of the haptoglobins of the two homozygous types (Bearn and Franklin).

In defining codominance and interference, we spoke of allelic genes which

FIG. 18.—Relative speed of migration of haptoglobins of different phenotypes in a starch gel at pH 8.4. The numbers indicate the hypothetical degree of polymerization. Anode marked by +.

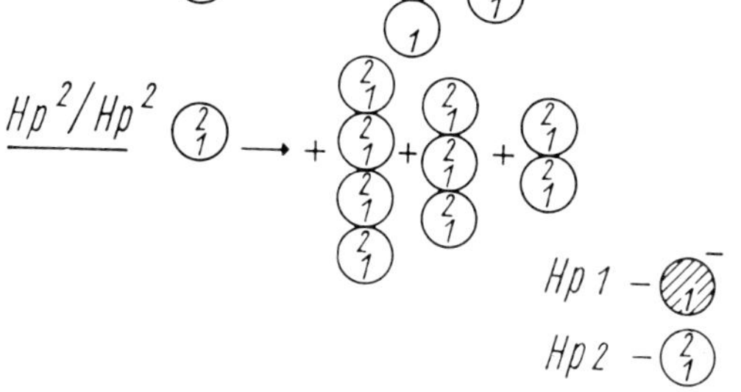

FIG. 19.—Hypothetical model of the polymerization of haptoglobin molecules and their relative electrophoretic speed of migration in starch gel. Only the first three polymers are shown. Anode on the right. (Polymer is used here in the chemical sense, as a giant molecule composed of similar constituent molecules, and not in the genetic sense of the word.)

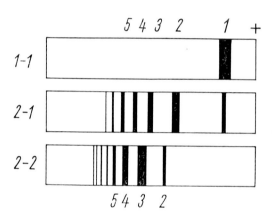

interact but whose individual effects are also known. Presumably there is, in addition, a special form of interaction of allelic genes which is not so easily analyzed, since the heterozygous action of only one of the two alleles is known, while the other manifests itself only by the fact that it modifies the heterozygous effect of its allele. This situation can be explained using the example of the sickle cell hemoglobin C disease, if we go back to the time when it was possible to diagnose only the sickle cell trait and sickle cell anemia, but when nothing was yet known about the possibilities of electrophoretic separation and identification of pathological hemoglobin types. At that time, persons who were heterozygous for gene Hb_β^C would not yet have been discovered; even though an atypical form of sickle cell anemia could have been diagnosed in the double heterozygotes $Hb_\beta^S \, Hb_\beta^C$, there would have been no possibility of specifying the diagnosis more closely. If we take another look at the family of Figure 17 from this old-fashioned point of view, it is remarkable that the siblings inherited an anomaly from their father which is more severe than his but of equal severity in them. This situation can be explained by the influencing of the dominant effect of a gene by the allele in the same locus. As a result of their relatedness, children and parents agree, on the average, in 50 per cent of their genes. Likewise, siblings agree, as a result of their relatedness, in 50 per cent of their genes (see p. 154). If, nevertheless, the agreement in the expression of a dominant disease is greater among siblings than between parents and children, this indicates that the "normal" allele exerts an influence. For it is in the normal allele that afflicted children often agree with one another but never with their afflicted parent, since, because of the separation of the alleles during the maturation division, two homologous alleles cannot be passed on to the same child. Up to now, the possibility of participation of the normal allele in dominant anomalies has not been taken into account. It is to be assumed from the start, however, that differences in the normal allele must be especially important for the manifestation of a dominant gene, since alleles affect the same circumscribed partial function. Penrose (1947–49) was the first to offer statistical arguments for the significance of the non-pathological allele for the manifestation of a dominant pathological gene, namely, in the case of myotonic dystrophy. In this disease, one finds muscular dystrophy primarily of the face and neck, myotonia of the hand muscles, psychic changes, hypogonadism, and cataracts. The appearance of the patients is often quite characteristic, with the limp position of the head, open mouth, and formation of baldness at the forehead. This disease exhibits the strange phenomenon of anticipation, which means the occurrence of a disease at a progressively younger age from generation to generation in afflicted families. In pre-Mendelian times, anticipation was regarded as a general law which was supposed to be valid for all hereditary diseases. It became evident, however, in the course of more thorough family investigations either that hereditary diseases show no anticipation or that anticipation is based on a selection effect. The reason for this is that only those members of the earlier generation are detected who married and had children and thus are selected for late onset of the disease. Myotonic dystrophy, if it occurs early, leads to deterioration of the personality, to social decline, and sterility. Presumably, the time and severity of incidence depend to a large extent on the constitution of the normal allele. At any rate, Penrose was able to show that siblings are in considerably greater agreement in regard to the age at which symptoms of the disease occur than are parents with their children. For the dominant hereditary nail-patella syndrome, it is likely that the dominant gene effect is influenced in a similar manner by the allelic gene.

Recessivity

A HOMOZYGOUS PAIR OF GENES DETERMINES
A CHARACTERISTIC. INHERITANCE FROM BOTH PARENTS

A gene which leads to a recognizable effect only in the homozygous state is called recessive. The homozygous state can occur only if both parents possess the gene in question. Thus, recessive inheritance is inheritance from both parents. If the more rare genes are involved, the probability is very small that a patient who is homozygous for a recessive gene will find a marriage partner who, by chance, has the same gene. As a result, the children of a patient with a recessively inheritable anomaly are usually free of the anomaly. But all of them have the trait in the heterozygous state, since, of course, they have inherited one or the other allele from their diseased parent. The parents of patients with recessive hereditary diseases also are usually healthy themselves. However, in the case of very rare recessive hereditary diseases, they are often blood relatives. The probability that a rare gene is present simultaneously in both marriage partners becomes considerably greater if the marriage partners have common ancestors and thus common genes. The probability that the germ cell of a heterozygous parent contains the gene is 0.5, and the probability for the mating of two such germ cells is $0.5 \times 0.5 = 0.25$. Therefore, on the average, 25 per cent of the children of heterozygous parents are homozygous. Since the number of children per family in Europe and the United States is usually small, most cases of recessive hereditary diseases occur sporadically. If the probability of becoming ill amounts to 0.25 for a child of heterozygous parents, then the probability that, in families with two children, both children are afflicted amounts to only 6.25 per cent $(0.25 \times 0.25 = 0.0625)$. For sixteen families with two children with heterozygous parents, one would expect one with two diseased children, six with only one diseased child, and nine with only healthy children $(1/16 + 6/16 + 9/16)$; this is calculated from the individual probabilities of $\frac{1}{4}$ and $\frac{3}{4}$ for healthy and diseased children according to the formula $(a + b)^2 = a^2 + 2ab + b^2$: $(\frac{1}{4})^2 + 2 \times \frac{1}{4} \times \frac{3}{4} + (\frac{3}{4})^2$. In families with three children, the frequency of families with three, two, one, and no diseased children is calculated by the formula $(a + b)^3 = a^3 + 3a^2b + 3ab^2 + b^3$, where, again, $a = \frac{1}{4}$, $b = \frac{3}{4}$, to be

> 3 diseased children 1/64
> 2 diseased children 9/64
> 1 diseased child 27/64
> no diseased child 27/64

If only families with two and three children occurred in the population, and these with equal frequency, the percentage of isolated, non-familial cases would be

$$\frac{6/8 \times 100 + 27/48 \times 100}{2} = 65.6 \text{ per cent.}$$

Thus, approximately 2/3 of all cases would occur as isolated instances. The above formula comes about because, in families with two children, for one family with two afflicted children there are six families with only one afflicted child, that is, 6/8 of the cases are isolated ones; also because, in families with three children, for one family with three cases there are nine families with two cases each—eighteen afflicted children—and twenty-seven families with only one afflicted child. Thus, twenty-seven isolated cases occur in a total of forty-eight cases $(3 + 18 + 27 = 48)$. Actually, of course, larger families also occur in the population, among which the chance that several children are diseased is higher; but there are also numerous families with one child. On the whole, therefore, the model does not deviate very far from the number of children per fertile marriage customary today, which usually lies between two and three. It is therefore understandable that, in most of the empirical investigations on recessive hereditary diseases, sporadic cases are more frequent than familial ones. Therefore, for any anomaly of unknown nature, a negative family anamnesis, even if it extends over many generations, can never exclude its genetic nature.

It is true that familial cases occur more often in the literature than in the population. Since it is customary especially for authors interested in genetics to publish such cases in the form of case histories, the familial occurrence is often an incentive for publication. Added to this is the fact that well-meaning colleagues communicate especially the familial cases to an author whose interest in genetics is known. For this and other reasons, case histories collected from the literature are not reliable material for statistical investigations in genetics.

The deductively derived theoretical frequency of patients with recessive hereditary diseases among the children of heterozygous parents gives a 1:3 ratio of diseased to healthy children if all such families can be detected—not only those in which diseased children actually occur but also those in which diseased children could, but accidentally did not, occur. However, since heterozygous parents usually can be recognized only by the fact that they have homozygous children, only those families are known in which the disease has actually occurred. In place of the distribution, calculated a priori, of healthy and diseased children in all families in which both parents are heterozygous, one therefore starts, in families having at least one afflicted child, with a truncated distribution. The theoretical distribution is $(a + b)^n$ for all families. In families containing patients, the term b^n is missing from the distribution, this being the term which corresponds to the number of families having only healthy children. The reduced distribution for families with at least one afflicted child thus becomes $(a + b)^n - b^n$. For families with two children, the result is $a^2 + 2ab$; for families with three children, $a^3 + 3a^2b + 3ab^2$; etc. The fraction of all families with the possibility of occurrence of homozygous children which is not included is as follows: for families with two children, $9/16 = (\frac{3}{4})^2$; for families with three children,

TABLE 9

NUMBER OF HEALTHY AND DISEASED CHILDREN
IN FAMILIES WITH THREE CHILDREN
WITH HETEROZYGOUS PARENTS

NUMBER OF FAMILIES	RATIO OF DISEASED TO HEALTHY	NUMBER		CORRECTED NUMBER	
		Diseased	Well	Diseased	Well
1.........	3:0	3	0	6	0
9.........	2:1	18	9	18	18
27.........	1:2	27	54	0	54
27.........	0:3	0	81	...	...
Total...........		48	144	...	...
Total (only families having diseased children).		48	63	24	72

$27/64 = (\frac{3}{4})^3$; for families with four children, $81/256 = (\frac{3}{4})^2$. In the example calculated above for families with three children, taking into account the families with healthy children results in the 1:3 ratio of the theoretical expectation, but including only families with diseased children gives a ratio of 1:1.31 (Table 9).

A calculation of the theoretical 1:3 ratio can be made also from data on families with at least one diseased child. Here, one starts with the consideration that the probability for the siblings of a homozygous patient of being homozygous themselves is independent of the genotype of the original case. Thus, if one counts only the siblings of the patients, but not the patients themselves, one also includes among the sets of siblings of the patients those which do not contain any further cases. One has thereby avoided the selection error which comes about if only sibships are counted which contain at least one afflicted member. In this method of counting, the patients serve, so to speak, only as indicators of families in which both parents are heterozygous. It is necessary here, however, to count the siblings of each patient, even if there are several patients in one set of siblings. Thus, in a family with three diseased and two healthy children, starting from each of the three patients two diseased and two healthy siblings apiece are counted, or a total of six diseased and six healthy children. This correction procedure has been carried out in the last column of Table 9. It can be seen that the theoretical 1:3 ratio is restored, even though only families with at least one patient have been included. This correction method originated with Weinberg. It is designated as the sibling method. The special conditions under which alone it may be applied will be discussed farther on.

The sibling method has the great disadvantage that information gets lost about the patients, who are the only diseased persons in their sibships. Since most cases of recessive hereditary disease occur sporadically, much of the

TABLE 10

AVERAGE NUMBER OF HOMOZYGOUS
CHILDREN IN FAMILIES WITH TWO
HETEROZYGOUS PARENTS AND AT
LEAST ONE HOMOZYGOUS CHILD

Number of Children per Family	Average Number of Homozygous Children per Family
2	1.14
3	1.30
4	1.46
5	1.64
6	1.83
7	2.02
8	2.22
9	2.43
10	2.65

total information about a disease is discarded. The number of patients on whom information is available is limited to those of families with several afflicted siblings. Thus, even for a relatively large number of propositi, the statistical error will remain undesirably large. The information contained in a collection of material may be utilized more completely if another procedure is followed. Essentially, this method was devised by Bernstein, improved in some critical points by F. Lenz (1929), and developed mathematically by Hogben. In the literature, therefore, it is sometimes designated as the a priori method of Bernstein-Lenz-Hogben. This designation is not quite appropriate, since Weinberg's sibling method is also based on comparison with a theoretically calculated "a priori" expectation. In detail, the procedure is as follows: The basis is again the truncated distribution which arises by not including families with exclusively healthy children. From this truncated distribution, one can calculate how many homozygous children are to be expected on the average in families with two, three, four, etc., children, if both parents are heterozygous. This yields the figures in Table 10.

The figures for the expectation of the number of homozygous children were calculated in the following way: In families with two children, for every family with two afflicted children, six families with only one afflicted child are to be expected. Thus, there are eight diseased children for every seven families, and $8/7 = 1.143$. In families with three children, for one family with three diseased children there are nine families with two and twenty-seven families with one diseased child. Thus, for thirty-seven families, there are $3 + 18 + 27 = 48$ diseased children, and $48/37 = 1.297$. Weinberg's sibling method as well as the "a priori" method of Bernstein-Lenz-Hogben presupposes that all families in the population with at least one diseased child are included, that is, that families with several diseased children are not included preferentially (see, however, p. 52). In order to understand the significance of this assumption, we must take a closer look at the conditions under which the material is usually collected. Here one can, in principle, distinguish two procedures, which we shall attempt to clarify by means of an illustrative comparison.

Let us imagine a pond, on the bottom of which there are flounders of diverse sizes. If we aim a fishing spear at random into this pond, we raise a "random sample" of flounders which, in the statistical sense, is not a perfect random sample, for the probability with which we hit a flounder by aiming blindly is proportional to its surface area. Therefore, we shall hit flounders which are four times as large as others with four times the frequency, if both are present in the same number. The "random sample" of flounders which we obtain with the fishing spear therefore does not agree in its size distribution with the flounder population which we would obtain by draining the entire pond and measuring and counting all flounders. For, in this case, larger flounders would not be recorded with greater frequency than small ones.

If one begins to collect individual cases in a hospital or in a medical practice with a relatively limited patient population, the procedure resembles the blind aiming with the fishing spear. A family with four diseased siblings here has four times the chance of a family with one case of being represented in our material. From the one case, we also reach the remaining three cases, which perhaps would never have become known to us without the initial case. Isolated cases do not have this possibility of being detected as secondary cases. Thus, in a record which starts from individual cases, the so-called probands, the truncated distribution is additionally distorted. Here, also, a correction is possible in principle. To be sure, it must be taken into account which were primary cases, or probands, and which cases became known to the investigator secondarily through the probands. Unfortunately, the distinction is not always simple, especially if careful records were not kept during the collection of the material as to the way in which the individual siblings were ascertained. It can happen that a patient is recorded as a proband and that his afflicted siblings are thereby labeled as secondary cases. It is possible, however, that the siblings might also be encountered with a high degree of probability at a later time, independent of the afflicted proband. In that case, they should really have been counted as new probands. If the probability of also including secondary cases independent of the probands approaches 100 per cent, the obtaining of material approaches the complete inclusion of the population. A proband method in which the other siblings are labeled as secondary cases as soon as the first probands of each sibship are recorded will, on the other hand, yield values which are too low.

In practice, the collection of data in human genetics will usually occupy an intermediate position between the two limiting cases of the sampling of individual probands and the gapless inclusion of all cases in the population. Often it is difficult to decide which extreme is approached more closely by this method. In general, however, gapless inclusion will be designated as such, so that, if specific data are lacking, ascertainment starting from single probands can be assumed. In the schematic representation below (Table 11), the two possibilities of including all cases in the population (1) and of inclusion starting with individual cases (2) are compared. The first case may also be defined so that the probability of independent inclusion of a secondary case equals 1 (100 per cent); for the second case, it is assumed that the probability of independent ascertainment of a secondary case equals 0. If it does

TABLE 11

DISTRIBUTION OF HEALTHY AND DISEASED SIBLINGS IN FAMILIES
WITH A SIMPLE RECESSIVE HEREDITARY DISEASE

1. Complete inclusion of all cases in a population limited in time and geography. "Truncated normal distribution."*			2. Sampling starting from individual cases. Selection of probands. Distorted truncated normal distribution.†		
Observed	Corrected		Observed	Corrected	
	Siblings Counted as:			Siblings Counted as:	
	Diseased	Healthy		Diseased	Healthy
In Families with Two Children					
● ●	2	...	■ ● ● ■	2	...
● ○	...	1	■ ○	...	1
● ○	...	1	■ ○	...	1
● ○	...	1	■ ○	...	1
○ ●	...	1	○ ■	...	1
○ ●	...	1	○ ■	...	1
○ ●	...	1	○ ■	...	1
Total..................	2	6		2	6

* In (1), each case must be counted separately, i.e., of three afflicted siblings, each has two afflicted siblings.
† In (2), only the siblings of each proband are counted, not the probands themselves.

● = carrier of the feature;   ■ = proband.

TABLE 11—*Continued*

In Families with Three Children

Pattern			Value	Value	Pattern			Value	Value
●	●	●	6	...	■●●	●■●	●●■	6	...
●	●	○	2	2	■●○	●■○		2	2
●	●	○	2	2	■●○	●■○		2	2
●	●	○	2	2	■●○	●■○		2	2
●	○	●	2	2	■○●	●○■		2	2
●	○	●	2	2	■○●	●○■		2	2
●	○	●	2	2	■○●	●○■		2	2
○	●	●	2	2	○■●	○●■		2	2
○	●	●	2	2	○■●	○●■		2	2
○	●	●	2	2	○■●	○●■		2	2
●	○	○	...	2	■○○			...	2
●	○	○	...	2	■○○			...	2
●	○	○	...	2	■○○			...	2
●	○	○	...	2	■○○			...	2
●	○	○	...	2	■○○			...	2
●	○	○	...	2	■○○			...	2
●	○	○	...	2	■○○			...	2
●	○	○	...	2	■○○			...	2
●	○	○	...	2	■○○			...	2
○	●	○	...	2	○■○			...	2
○	●	○	...	2	○■○			...	2
○	●	○	...	2	○■○			...	2
○	●	○	...	2	○■○			...	2
○	●	○	...	2	○■○			...	2
○	●	○	...	2	○■○			...	2
○	●	○	...	2	○■○			...	2
○	●	○	...	2	○■○			...	2
○	●	○	...	2	○■○			...	2
○	○	●	...	2	○○■			...	2
○	○	●	...	2	○○■			...	2
○	○	●	...	2	○○■			...	2
○	○	●	...	2	○○■			...	2
○	○	●	...	2	○○■			...	1
○	○	●	...	2	○○■			...	2
○	○	●	...	2	○○■			...	2
○	○	●	...	2	○○■			...	2
Total..................			24	72				24	72

happen by the second method that two afflicted siblings are included as probands, both must be counted. In the second method, the probands themselves are counted only if there is a second proband in the family. In both cases, one finds the theoretical 1:3 ratio of diseased to healthy persons, if the conditions of selection really agree with those assumed in the calculation.

In the ideal case for genetic statistical calculations, namely, the complete inclusion of all cases in the population, reliable calculations are possible. It is necessary only to check carefully whether all cases have really been included. With an ailment which leads to early death, such as is the case, for example, in cystic fibrosis of the pancreas, special attention must be paid to the age limit placed on the population. Baumann has carried out genetic statistical investigations on sixty-three children having this ailment, of whom only fourteen reached an age above twenty-four months. Nine previously deceased cases were detected subsequently through anamnesis, by way of later siblings who were under clinical observation; they were included in the calculation. Although it appears in this investigation that practically all cases in a certain region of Switzerland between 1945 and 1957 were included, the total material cannot be considered free of selection. Using Bernstein's a priori method, Baumann found 72 afflicted children compared to an expected 57.2. The difference was "statistically significant." It was concluded from this that, presumably, some heterozygous siblings become ill and that cystic fibrosis of the pancreas thus is not a simple recessive hereditary disease. It appears to be more likely that isolated cases of early death were not included as completely as familial ones which could still be traced by way of later siblings.

In cystic fibrosis of the pancreas and several other recessive hereditary diseases, an attempt has been made to explain a percentage of diseased siblings which appeared to be too high for simple recessivity by saying that some heterozygous siblings also become ill. This explanation, however, would be valid only if heterozygous carriers of the trait occurred exclusively in families with homozygous patients. In reality, most heterozygous carriers of the gene occur in families which do not contain any homozygotes. Thus, if individual heterozygotes become ill, the occurrence of isolated cases would increase. Then the total percentage of diseased persons among the siblings of patients would not be higher, but rather lower than expected in simple recessive heredity.

Regardless of whether Weinberg's sibling or proband method or the Bernstein-Lenz-Hogben a priori method is applied, the way in which the material was obtained must always be taken into consideration—whether it was by complete inclusion of all families or by limited random sampling of individual probands. The expectation figures of Table 10 are based on the assumption that all sibships containing patients were included. In order to obtain comparable figures by the selection of probands, it would be necessary, according to a suggestion by F. Lenz (1929), to count a set of siblings with n probands n times, but then to divide the number obtained by the number of all afflicted siblings. For example, a family with three carriers of the trait among six siblings, of whom only two are probands, would contribute $3/3 \times 2 = 2$ diseased persons and $3/3 \times 2 = 2$ healthy siblings.

In the case of material obtained in a manner that cannot be determined exactly, the calculation can be carried out in two ways, namely, by assuming that there was complete inclusion, and by assuming that individual probands formed the point of departure. This double procedure can be used, for example, with data collected from the literature or with material from case histories which contain precise family anamneses but which make no distinction between probands and secondary cases. Two different values will then be found, between which the theoretical value of 0.25 should lie if the trait is recessive. However, the disparity between the results of the two methods usually is so great that one cannot obtain a very precise answer. Thus, in a series of ninety-three families with cystic fibrosis of the pancreas, a value of 0.29 was found with the assumption of complete inclusion, and 0.18 with the assumption of inclusion proportional to the number of patients per family. For material from the literature, consisting of eighty families with Gaucher's disease, the values were 0.30 and 0.21.

The main difficulty in giving a clear demonstration of the methods for examining the recessive mode of inheritance lies not so much in the determination of the proportion of afflicted and non-afflicted siblings but in the calculation of the statistical significance. The most advanced textbooks shy away from explaining the calculation of the statistical error (Fraser Roberts). The customary calculation of errors, which is based on a normal distribution that has not been truncated, cannot be considered. In the striving for the most complete method, electronic computers or voluminous tables have been used as an aid for evaluating the significance. The specific literature on mathematical statistics for testing hypotheses on modes of inheritance, and for estimating frequencies among siblings of patients, is rather extensive. A discussion is not possible within the framework of this book, but interested readers are referred to several important publications (Bailey, 1951a, 1951b; Finney; Haldane, 1932, 1937; Hogben; Lejeune, 1958; Smith, 1959). The entire problem is discussed extensively by Kaelin, whose tables are also highly recommended for practical use.

With the material which the physician usually has at hand, namely, a relatively limited series of cases of his own or a series collected from the literature, one can hardly expect a practical advantage in the application of the more complete modern methods as compared to the older crude procedures. For special genetic investigations, collaboration with a statistician interested in genetics is necessary. The participating physician should also have enough insight into the statistical posing of the problem so that no misunderstandings occur during the collaboration. Quite generally, a warning is needed that not too much significance should be attached to presumable deviations of observation from expectation, even if they should be "statistically significant." Errors in obtaining the material, which cannot be removed by any subsequent statistical calculation, are by far the most frequent cause of such discrepancies.

An important criterion for recessive inheritance is the observation that only diseased children can come from a marriage between two patients having the same recessive hereditary disease, since both parents are homozygous and can thus provide only one type of allele. The emphasis here, however, is on the sameness of the recessive hereditary disease; for, if two patients marry who have a disease which appears identical externally and is inherited recessively by both, but which is determined in each by a different, non-allelic gene, then all children, though heterozygous carriers of each of the two genes, are phenotypically healthy. In certain circumstances, the observation of healthy children from marriages between patients with what appears to be the same hereditary disease is the only indication that a hereditary disease is genetically not homogeneous, but heterogeneous. The analysis of such cases has attained special significance for the genetic study of hereditary deaf-mutism. Deaf-mutes often intermarry. All children of such a marriage are often also deaf-mutes. In the majority of cases, however, all children are healthy, and this is true even if only those marriages are taken into account in which both partners obviously suffer from hereditary deaf-mutism. Evidently, there is a series of different genes which lead to deaf-mutism in the homozygous state. In Figure 20, a pedigree is reproduced in which both types

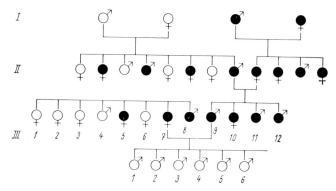

FIG. 20.—Pedigree with recessive deaf-mutism. Two marriages between deaf-mutes produced only deaf-mute children. Evidently, the same genetic type was present here. A third marriage between deaf-mutes produced only healthy children. This was evidently a case of two different types, corresponding to the chromosome scheme. (From Stevenson and Cheeseman.)

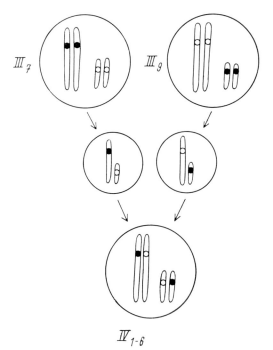

FIG. 21.—Chromosome scheme for the pedigree above

of marriages between deaf-mute partners occur side by side. In a family described by Trevor Roper, both parents were albinos with pink iris and albinotic fundus, strabismus, and nystagmus, but all three children had normal pigmentation. If the legal father was identical with the biological one in this case, one must assume that there are two different genetic forms of albinism.

Usually, recessive hereditary diseases occur only in one generation, that is, they are "familial" according to pre-Mendelian nomenclature in contrast to the "hereditary," dominantly inherited, diseases which occur through several generations. Under special circumstances, however, a recessive hereditary disease can occasionally be traced through several generations. The pedigree in Figure 20 gives an example of the occurrence of a recessive anomaly in three generations. This behavior, which is unusual for recessive anomalies, can be explained by the fact that deaf-mutes preferentially intermarry, so that the result is, in a sense, a pure breeding of the gene. But a similar picture of heredity through several generations can occur if a pathogenic recessive gene has become widely distributed throughout an area of inbreeding. In such a case, the gene may become homozygous repeatedly in several successive generations. This is evidently the basis of a pedigree of presumably dominant alkaptonuria, in which the anomaly occurred through four generations, and this in a rural Spanish-Indian hybrid family on Santo Domingo Island, a family in which consanguineous marriages were frequent. The fact that such pseudodominance of recessive hereditary diseases is not observed more frequently is due primarily to the severity of a large number of recessive hereditary diseases, which prevents the patients from reaching adulthood or marrying. In any case, the appearance of a rare anomaly in three or more successive generations is a reliable criterion of dominant heredity only in the absence of consanguineous marriages.

SIGNIFICANCE OF CONSANGUINEOUS MARRIAGES

One of the decisive criteria for recessive hereditary causation of rare diseases is the increased frequency of marriages between blood relatives among the parents of patients. In order to understand this, we must start with some considerations regarding the frequency of recessive genes in the population. Ordinarily, a patient who is homozygous for a pathological gene has two heterozygous parents; for each homozygous sibling he has two who are heterozygous; he has only heterozygous children. The frequency of heterozygous carriers of genes is many times, often several hundred times, higher than that of patients. In the heterozygous state, recessive genes may be inherited through numerous generations—on the average, presumably through several hundred generations, until they accidentally meet a like allele and thus become homozygous. If matings occur purely by chance in the population ("panmixia"), then evidently the possibility that a recessive gene will meet a gene of the same kind depends merely on the frequency of this gene in the population. Here, also, the probability of coincidence of two events is equal to the product of the probabilities of the individual events. If a certain gene represents 1 per cent of all alleles of the same series in the entire population, the probability for a random gene of this series to be that specific gene is 1 per cent or, as it is usually written to simplify the frequency calculation, 0.01 (1 = 100 per cent). The probability of coincidence of two such genes amounts to $0.01 \times 0.01 = 0.0001$ $(1/100 \times 1/100 = 1/10,000)$. Or, expressed more generally, if the gene frequency within the allele series amounts to p, then, under the assumption of panmixia, the frequency of homozygous individuals in the population amounts to p^2. If it is assumed that the gene locus can be occupied only by either the pathological gene or a normal allele, the frequency of the normal allele amounts to $q = 1 - p = 0.99$ (or 99 per cent). Now the frequency of heterozygous individuals is calculated from the binomial $(p + q)^2 = p^2 + 2pq + q^2$, where p^2 represents the frequency of homozygotes of the pathological allele, q^2 the frequency of homozygotes of the normal allele, and $2pq$, the frequency of heterozygotes. For a gene with a frequency of 0.01, we obtain a heterozygote frequency of 0.0198 (or almost 2 per cent), that is, practically twice the gene frequency. The lower the gene frequency, the more closely the heterozygote frequency approaches twice the gene frequency. The frequency of homozygotes in our example is 0.0001. Thus, for one homozygous carrier of a gene there are almost two hundred heterozygous carriers.

In this example, we started out purely by deduction from a theoretical gene frequency. In recessive hereditary diseases, however, the gene frequency is at first unknown; only the homozygote frequency is known. Under the hypothesis of random matings, it is possible to calculate from the homozygote frequency p^2 the gene frequency p, and from it, in turn, the heterozygote frequency $2pq$. Since $q = 1 - p$, the heterozygote frequency is $2(p - p^2)$, or, in our example, $2 \times (0.01 - 0.0001) = 0.0198$. The relationship between homozygous patients and heterozygous carriers of a gene will be shown once more by means of an arithmetically favorable example, namely, that of cystic fibrosis of the pancreas. This is a generalized disease of the exocrine glands in which insufficient pancreatic digestion, increased tendency toward pneumonias with development of bronchiectases and emphysema, and an increase in the sodium and chloride content of the sweat are clinically predominant. The frequency is between 0.7 and 1 per 1,000 live births. To simplify the calculation, 9 : 10,000 is assumed as the correct value. This value corresponds to p^2, the square of the gene frequency. The gene frequency turns out to be $\sqrt{9 : 10,000} = 3/100$, or 3 per cent. The frequency of heterozygotes would then be $2(p - p^2) = 5.82$ per cent, or 0.0582. Thus, if not quite every thousandth person in the population has a recessive disease, no less than approximately every seventeenth (5.82 out of 100) possesses the gene in the heterozygous state.

The considerations thus far regarding the relative frequency of patients and carriers of genes are valid for panmixia. In the actual population, panmixia is never completely attained, but there is a more or less pronounced degree of homogamy, or assortative mating; that is, among similar families, geographical areas, social strata, religious communities, etc., marriages occur more frequently than would be expected in purely random mating. Because of this homogamy, genes inherited from common ancestors meet more frequently than would be expected in panmixia. To be sure, only marriages between close blood relatives are of significance here, chiefly those between cousins. On the average, as a result of their relatedness, siblings have half their genes in common, as do parents and children. The probability that a rare gene is present in common in two cousins therefore amounts to $\frac{1}{2} \times \frac{1}{2} \times \frac{1}{2} = \frac{1}{8}$. If, as in the first sample calculation, a random individual is heterozygous for a certain gene with a probability of 0.0198, the same probability of 0.0198 would be valid for his non-related marriage partner. The combined probability that both partners in a random marriage are heterozygous would amount to $0.0198 \times 0.0198 = 0.000392$, or approximately 0.04 per cent. In contrast, in a cousin marriage the probability that both partners have inherited a certain gene of frequency 0.01 from their common grandparents amounts to 0.00247 ($\frac{1}{8} \times 0.0198$), so that it is more than six times as great as the general probability that two random marriage partners are heterozygous for this gene. Added to this is the fact that, even in the $\frac{7}{8}$ of cousin marriages in which both partners have not inherited the common gene, the pathological gene can be present in both by chance and independently of the fact of blood relationship. For this $\frac{7}{8}$, the same probability holds as for non-related marriage partners that both are heterozygous. Thus, $\frac{7}{8} \times 0.000392$ has to be added to 0.00247. The total chance, for a gene frequency of 0.01, that both partners in a cousin marriage are heterozygous is then 0.00286, or 7.6 times as great as for random, non-related marriage partners.

If one now wants to know how frequently consanguinity is to be expected among the parents of patients with a certain recessive hereditary disease, the frequency of consanguineous marriages in the population must be taken into account. For a frequency of cousin marriages of 1 per cent, there would be $0.01 \times 0.00286 = 0.0000286$ cousin marriages for $0.99 \times 0.000392 + 0.01 \times 0.00286 = 0.0004146$ total marriages of parents of patients (non-blood relatives + blood relatives), so that 6.9 per cent of the patients would be children of blood-related parents.

For the above example of cystic fibrosis of the pancreas, we shall now calculate the percentage of cousin marriages which is to be expected. As we have seen, the heterozygote frequency amounts to 0.0582 ($p = 0.03$, $2p - 2p^2 = 0.0582$). The probability that both partners in a random marriage are heterozygous for the gene amounts to $0.0582 \times 0.0582 = 0.0039$, or approximately 0.34 per cent. Since only $\frac{1}{4}$ of the children from such marriages become homozygous, we obtain $\frac{1}{4} \times 0.34$ per cent = approximately 0.085 per cent homozygous in the population. The difference between 0.0847 per cent and 0.09 per cent shows up an error in our model which will concern us later on, in connection with the discussion of the concept of mutation. Since patients with cystic fibrosis of the pancreas do not propagate, about 6 per cent of all genes become victims of elimination in each generation: for one homozygous patient with whose death two genes are eliminated there are 32.3 genes in heterozygous carriers of the gene. Thus, to begin with, there is no equilibrium in the population. One would expect that the genes for a hereditary disease which is so frequently lethal would disappear relatively quickly from the population. Evidently the loss of genes is either compensated for in each generation by new mutations, or the heterozygous carriers of the gene must possess some kind of advantage which assists the spreading of the gene. We shall first calculate the probability that both partners in a cousin marriage are heterozygous carriers of the gene. The probability that the gene is present in the heterozygous state in one of the partners amounts to 0.0582; the probability that the other partner has inherited the same gene amounts to $0.0582 \times \frac{1}{8} = 0.00728$. Added to this is the probability that both marriage partners possess the same gene by chance; this amounts to $\frac{7}{8} \times 0.00339 = 0.00297$. Altogether, then, of all cousin marriages, $0.00728 + 0.00297 = 0.01025$ are between heterozygotes. The frequency of consanguineous marriages among the parents of patients now depends on the percentage of consanguineous marriages in the population. If we assumed a frequency of 1 per cent, we would find in the population 0.01×0.01025 consanguineous marriages between heterozygotes for 0.99×0.00339 marriages of heterozygotes who are not blood relatives, or $0.0001025 : 0.00336 = 1 : 32.8$. Thus, the percentage of consanguineous marriages among all families with diseased children would be

$$\frac{0.0001025}{0.0001025 + 0.00336} = \frac{0.0001025}{0.00346} = 3.37 \text{ per cent}.$$

For an anomaly which is as frequent as cystic fibrosis of the pancreas, therefore, one can expect only a threefold increase in the frequency of blood-related parents. This increase usually cannot be detected statistically. The mean error of a percentage is calculated according to the formula $\sqrt{x \cdot y/n}$. In our case, $x = 3.37$, $y = 1 - x = 96.63$; for 100 cases, the mean error would be 1.79. The difference between the expected percentage of 3.37 and the percentage of consanguineous marriages in the population amounts to 2.37 per cent; that is, it would not be twice as large as the mean error and therefore not statistically significant. For 200 cases, the mean error would be 1.28, so that the difference is still not certain; only at 300 would the mean error amount to 0.9, so that the difference would exceed twice the mean error. In fact, in the case of cystic fibrosis of the pancreas it has not yet been possible to prove any increased consanguinity of the parents. For such a common disease, this is no argument against recessive inheritance. The more rare a recessive gene, the higher the percentage of consanguineous parents. In the limiting case where a gene was produced only a single time in a family through mutation and was inherited by only a few descendants in the following generations, this gene could become homozygous only through consanguineous marriages. Here, then, 100 per cent of all cases came from consanguineous parents. If, on the other hand, a gene is so widespread in the population that the chance is high even for random, non-related marriage partners that both are heterozygotes, the blood relationship plays a small role. For a gene frequency of 0.1 the frequency of the recessive anomaly would be 0.01 (1 per cent), and the heterozygote frequency, 0.18 ($p = 0.1$, $2p - 2p^2 = 0.18$). If the frequency of cousin marriages in the population again were 1 per cent, then 1.56 per cent of the parents of patients would be blood relatives compared to 1 per cent of random parents. Such a small difference cannot usually be detected even under the best conditions of obtaining material, especially since the frequency of consanguineous marriages is not constant but exhibits considerable fluctuation according to social position, city or country population, and geographical and temporal origin. In

TABLE 12

FREQUENCY OF MARRIAGES BETWEEN FIRST-DEGREE COUSINS

Region	Author	In Per Cent of All Marriages
Berlin, 1896–1913	Czellitzer	0.57
Württemberg (county)	Spindler, 1925	1.8
Netherlands, 1936–41	Official statistics	0.21
Netherlands, 1948–53	Official statistics	0.13
New York, 18th century	G. H. Darwin	2.76
U.S.A., present	Glass	0.05
Bavaria	Wulz, 1925	0.6
Bornholm	Strömgren, 1938	1.3
Copenhagen	Bartels, 1941	1.2

modern metropolitan populations, consanguineous marriages have become very rare. The decrease in the number of children and the increasing intermingling of the population are responsible for this. Today, a young man generally has a choice of fewer cousins and more non-related girls than did a village resident in the nineteenth century. Table 12 gives some data on the frequency of first-degree cousin marriages.

In some regions of inbreeding, especially in Japan but also in Switzerland and in Scandinavia, the percentage was considerably higher, but in these regions, also, inbreeding is gradually disappearing from one year to the next (Obermatt, Switzerland, 11.5 per cent [Egenter]; Muonionulusta, Sweden, 6.8 per cent [Böök, 1948, 1957]; Nagasaki, Japan, 8.0 per cent [Schull, 1953]).

The only demonstrated effect of consanguineous marriages consists in the fact that they increase the risk that recessive genes will become homozygous. Altogether, this risk is not very high even for children from consanguineous marriages. However, a larger number of malformations and more deaths in childhood occur among children from cousin marriages.

The percentage of consanguineous marriages among the parents of patients is not a biological constant. In addition to the gene frequency, it depends on the frequency of consanguineous marriages in the population. The gene frequency is kept relatively constant by the mutation rate and by selection, but the frequency of consanguineous marriages in the general population is subject to large fluctuations. In the more modern metropolitan population, consanguineous marriages are so rare that their occurrence has lost its usefulness as a criterion of recessive inheritance. In the metropolis of New York, among 102 families in which children with amaurotic idiocy of the Tay-Sachs type were born, the parents were blood relatives in only two cases. In some older statistics, for Jewish patients with infantile amaurotic idiocy, consanguinity of the parents was found in 23 per cent (Slome, pub-

TABLE 13

FREQUENCY OF RECESSIVE HEREDITARY DISEASES AND PERCENTAGE
OF CASES FROM CONSANGUINEOUS MARRIAGES

Disease	Frequency of Homozygotes	Frequency of Heterozygotes	Consanguineous Marriages
Cystic fibrosis of pancreas (mucoviscidosis)	0.0009	0.06	no demonstrable increase*
Adrenogenital syndrome.	0.0002	0.028	no demonstrable increase†
Infantile amaurotic idiocy (Tay-Sachs)‡..	0.00012	0.022	2%
Albinism.	0.00005	0.014	8%
Phenylketonuria.	0.000035	0.012	10–12.5%
Cystinosis.	0.000025	0.01	12%
Pfaundler-Hurler disease (autosomal recessive form).	very rare	. . . . .	20–30%
Chediak-Higashi-Steinbrinck anomaly§....	a total of 9 cases known, 5 of them from consanguineous marriages		

* Among 455 families in the literature, 1 cousin marriage and 6 more remote marriages between relatives.

† In the literature, isolated cases from consanguineous marriages.

‡ Frequency among Jews in New York. In other populations, considerably lower frequency.

§ Syndrome with albinism, tendency toward septic conditions in infancy. Lack of resistance, hepatosplenomegaly with icterus, enormously large granules of the white blood corpuscles.

lications of 1884–1933) and 18.6 per cent (Klein and Kténides, publications of 1934–54). In the non-Jewish population, the frequency of infantile amaurotic idiocy was calculated to be 0.0000022, and that of heterozygotes, 0.0028. The parents of the non-Jewish patients were related in 55.6 per cent (Slome) and 33.3 per cent (Klein and Kténides) of the cases.

The criterion of consanguineous marriages is useful especially for evaluating very rare anomalies where not enough cases are available for a genetic statistical analysis. In reports of case histories of rare anomalies, it should always be indicated whether the parents are related. In the Ellis–van Creveld syndrome, which includes six fingers on the hands, dental anomalies, short limbs, and fused bones in the base of the hand, it was possible to suspect in two of the first three known cases that the condition was due to recessive inheritance, on the basis of consanguinity of the parents, before its occurrence in the family had been observed. In the apparently very rare Berardinelli syndrome, only nine cases of which are known with certainty, again blood relationship of the parents in two families is a critical indication for recessive hereditary determination. The Berardinelli syndrome refers to a peculiar anomaly of the metabolism and of the endocrine glands, in which the patients remind one in part of lipodystrophy, in part of an adrenogenital syndrome with muscular hypertrophy and large penis or large clitoris; but, in addition, they exhibit acromegalic features and limbs and hepatospleno-megaly as well as hyperlipemia. Also, among the very isolated cases of lipoid hyperplasia of the adrenal cortex with male pseudohermaphroditism which have become known, a few came from related parents, so that recessive hered-ity can be assumed as the cause.

Not infrequently, the physician is presented with the question whether a consanguineous marriage should be advised against. The occurrence of a recessive hereditary disease among the close relatives is a weighty reason for deciding against marriage between blood relatives. Regularly dominant anomalies in the family of which the marriage candidates themselves are free, have no significance for the risk of a consanguineous marriage; the same is true for recessive sex-linked anomalies. If related marriage candidates share any small, apparently harmless morphological anomaly, this is a serious reason against marriage, since it is possible that the responsible gene in the homozygous state can lead to severe malformations in the children. Even if the families of both marriage partners have been free of hereditary diseases through generations, there is unfortunately no guarantee that only healthy children will be produced from a consanguineous marriage. Rare recessive hereditary diseases usually occur only in a single set of siblings with-in the wider family circle. Every observant physician has seen numerous cases in the course of the years in which consanguineous marriages have led to severe chronic ailments or early death of the children. Therefore the physician, if he is asked, should not keep silent about the fact that every consanguineous marriage means an increased risk.

Attempts have been made to determine the order of magnitude of the risk of a cousin marriage by means of extensive and systematic investigations. In the fre-quency of miscarriages and stillbirths, no difference between related and non-related marriages was found. Also, early death of the ovum, before a recognizable abortion can occur, is probably no more frequent in cousin marriages. At least the intervals between marriage and birth of the first child were the same for re-lated and non-related marriages. Unrecognized early lethality of the fetus would have to show up statistically in a lengthening of this interval. The mortality of children from consanguineous marriages was considerably higher in some in-vestigations than that of children from non-consanguineous marriages. The result of a French study was that, of 461 children from cousin marriages, 115 (25 per cent) died before they reached adulthood; but of 1,628 children from control fami-lies, only 210 (13 per cent) died. In an American study, 13 (6.3 per cent) among 205 children from consanguineous marriages died in childhood or youth, while only a single child among 164 children from control families died. According to data from Japan, among 352 children from cousin marriages, 41 (12 per cent) had died up to the ninth year of life, compared to only 31 (5.5 per cent) out of

567 children of non-related parents. In the same study, the frequency of severe congenital malformations among 2,846 children from cousin marriages was distinctly higher, 1.7 per cent, than for 63,796 children used for comparison, 1.0 per cent. The difference between the two groups was due primarily to the greater frequency of multiple malformations among the children of blood-related parents.

These investigations showed that a great many conditions of illness occur more frequently among the children of related parents than among other children, but, surprisingly, known recessive hereditary diseases are hardly involved in this difference.

If the increased mortality of children from consanguineous marriages were determined only by recessive genes which become homozygous, the frequency of this "lethal equivalent" in the population could be calculated theoretically. The concept of the "lethal equivalent" here includes genes which always lead to the death of the individual in the homozygous state, as well as those which lead to death only with a certain probability. A gene which would lead in the homozygous state to the death of 50 per cent of its carriers would be considered equal to 0.5 lethal equivalents. The calculation of the lethal equivalent has been carried out in many cases on the basis of the mortality of children of related parents. In the course of this, two to ten lethal equivalents per person were found.

An important defect in all studies of children from consanguineous marriages, however, makes one regard these figures with reservations. Consanguineous marriages result preferentially in families with many children. But, for economic and sociological reasons, there is such a distinct correlation between the number of children per family and the mortality during the entire period of childhood that the mortality of children from consanguineous marriages must also be affected by this. Added to this is the fact that, in some countries, consanguineous marriages occur preferentially in socially unaccepted kindreds whose members have difficulty in finding marriage partners from the remaining population. Also, the men who enter into consanguineous marriages are more often weak in making contacts, or are of low intellectual capacity. Even though the calculation of the lethal equivalent is interesting from a theoretical viewpoint, and though it may be useful as an exercise in thinking about population genetics, here again the complexity of the human social structure makes the result questionable.

The risk of consanguineous marriage diminishes rapidly with the degree of relatedness. Marriages between cousins and between uncles and nieces involve a risk which is small, but nevertheless to be considered. Among second cousins, the danger that children will become homozygous for rare recessive genes is already distinctly smaller. If one partner in a marriage between second cousins is heterozygous for a rare recessive gene, the probability that the marriage partner has the same gene is only 1/32. The total risk that a child will be homozygous for this gene amounts to only $1/4 \times 1/32 =$ 1:128, so that it lies within the order of magnitude of the general risk of severe malformation for a random child.

Blood relationships among grandparents and further antecedents are without significance for health. By means of consanguineous marriages, genes become homozygous only in a single generation; in the next, they again divide regularly. One can even say that a person in whose family numerous consanguineous marriages have occurred in earlier generations, and where these have not led to recessive hereditary diseases, has already passed a test, although an incomplete one, of his hereditary material for recessive genes.

THE DIFFERENT DEGREES OF RECESSIVITY. PARTIAL
MANIFESTATION OF HETEROZYGOTES

We have seen that dominance and recessivity are complementary concepts. If a gene is dominant, the corresponding allele is recessive. This is true first of all for completely dominant genes which cannot be distinguished by phenotype in the heterozygous and homozygous states. It follows from this that the allele of such a dominant gene has no recognizable phenic effect in the heterozygous state; it is recessive. The complementary nature of dominance and recessivity is valid also for the usage common in medicine, according to which a gene is called dominant if it produces pathological phenomena in the heterozygous state, even if the homozygous state leads to more severe phenomena. An incompletely recessive allele is complementary to an incompletely dominant gene. The nomenclature becomes difficult if a gene with intermediate dominance determines a phenotype in the heterozygous state, which is located midway between the phenotypes determined by the two alleles in their homozygous states. In such a case, one can call one gene dominant with the same justification as the other. Here we are confronted with the same problem of the relativity of dominance and recessivity which we encountered in the discussion of the dominant gene effect. For the practical purposes of medicine, we start with the pathological phenotype. If the intermediate state is of pathological significance, it is practical to regard the responsible gene as having intermediate dominance; but if the intermediate state can be uncovered only by means of refined methods of investigation and only the homozygous state has pathological significance, the gene is called recessive. As in dominance, different degrees of recessivity can be distinguished:

Complete recessivity: The gene can be detected only in the homozygous state.

Incomplete recessivity: The gene produces pathological symptoms only in the homozygous state, but can be detected in the heterozygous state.

Some recessive genes can never be detected in the heterozygous state, some only irregularly and only by special methods of investigation; but others regularly cause distinct deviations which cannot quite be designated as pathological. In the last case, the decision whether to speak of dominance or recessivity is especially difficult. In the compilation of anomalies with intermediate dominance on pages 38 ff., several anomalies are cited which could also be designated as recessive with partial manifestation in the heterozygous state. Thus, Böök started with a chondrodystrophic child who was being observed in an orthopedic hospital, and found only upon investigating the family that both parents were small and short-limbed and that this peculiarity of habitus could be traced as a dominant feature through four generations of the family. In contrast, Mohr and Wriedt first studied the dominant hereditary trait of brachymesophalangy of the second fingers in a Norwegian family. Only secondarily did they come upon a probably homozygous child from a cousin marriage between two afflicted family members; all of the child's fingers and toes were missing. If this child had been the starting case, and if the brachymesophalangy had been discovered only secondarily among the members of the family, it would have been more obvious to speak of a recessive gene with harmless partial manifestation in the heterozygous state.

If a gene appears to be completely recessive, this is presumably due only to the incompleteness of our morphological observation and chemical investigation of the phenotype. If we were morphologically all-seeing and chemically omniscient, we probably would always be able to decipher the genotype from the phenotype. At any rate, one must always realize that the concepts of dominance and recessivity depend on the ability to recognize the differences between the heterozygous and homozygous states. That no differences can be recognized does not prove that none are present. As our methods of investigation improve, the purely recessive genes become more and more rare. In the last few years, it has been possible in the case of numerous recessive hereditary diseases to determine small deviations even in the heterozygous carriers of the gene. Table 14 gives a list of recessive hereditary diseases with partial manifestation of the heterozygous state.

The recessive anomalies cited have the characteristic in common that they can be traced back to the absence of a gene product, usually an enzyme, and that, in the heterozygous state, frequently a decrease of the gene product or a deviation in the metabolism can de demonstrated, such as would be expected as a result of a decrease in the gene product. Here the heterozygous state does not lead to signs of illness. As a rule, a single normal allele evidently guarantees sufficient enzyme production. Normally, the organism operates with a margin of safety, so that the diminution of an enzyme by 50 per cent does not result in a disturbance of the metabolism. Considered functionally, therefore, the difference between the heterozygous and the homozygous normal state is usually meaningless, while the difference between the two on the one hand and the homozygous pathological state on the other hand can mean the decision between life and death, although, regarded purely quantitatively, the heterozygous state can be almost exactly intermediate. In the adrenogenital syndrome with congenital adrenal-cortical hyperplasia and pseudohermaphroditism in girls, and pseudopubertas praecox in boys, experiments to demonstrate the existence of heterozygotes so far have not led to conclusive results. In the rare cases of chronic coproporphyria with inhibition of growth, late rickets, hyper-aminoaciduria, and central cataract, increased elimination of coproporphyrin in the stools and urine as well as hyper-aminoaciduria can be shown in the heterozygotes also.

In cases of sphingolipidosis, which includes the three types of Gaucher's

TABLE 14

RECESSIVE ANOMALIES WITH PARTIAL MANIFESTATION IN THE HETEROZYGOUS STATE

Anomaly	Homozygous State	Heterozygous State
Afribrinogenemia	No demonstrable fibrinogen in the plasma; blood does not coagulate even after standing for several weeks; severe bleeding often leads to death in childhood	Occasionally decrease in fibrinogen without clinical symptoms
Acatalasemia	Absence of catalase activity in peripheral blood; upon addition of H_2O_2, blood at once turns black; gangrenous ulcers of the mucosa	Decrease of catalase to values outside the range of variation of normal controls; tendency to alveolar pyorrhea?
Cystic fibrosis of the pancreas	Digestive insufficiency due to decrease of pancreatic enzymes, tendency to bronchopneumonia and bronchiectases, cirrhosis of the liver, meconium ileus of the newborn; Na and Cl content in sweat usually above 50 meq/l	Approximately half of the heterozygotes have abnormally high Na content (above 50 meq/l) in sweat, also increased Cl content; tendency toward emphysema in old age?
Galactosemia	Icterus, starting in the first days of life; decrease in weight, hepatomegaly (cirrhosis of liver), later splenomegaly, cataract in the first weeks of life, feeble-mindedness, diminished resistance to infection, galactosuria, albuminuria, hyperaminoaciduria; deficiency of galactose-1-phosphate-uridyl transferase in erythrocytes and liver; abnormal curve of galactose content in blood (test is dangerous!)	Diminished content of galactose-1-phosphate-uridyl transferase in erythrocytes. Abnormal curve of galactose content.
Glycogen storage disease of liver and kidneys (von Gierke)	Absence of glucose-6-phosphatase in the liver cells; storage of glycogen in different organs, hepatomegaly, stunted growth, tendency to hypoglycemia; absence of hyperglycemia upon adrenaline administration; accumulation of glucose-6-phosphate and fructose-6-phosphate in the erythrocytes	Accumulation of glucose-6-phosphate and fructose-6-phosphate in the erythrocytes; diminished rise of blood sugar upon administration of adrenaline; no clinical symptoms
Hypophosphatasia	Congenital, rickets-like disease of the skeleton with hypercalcemia and tendency toward nephrocalcinosis; alkaline phosphatase reduced to a tenth to a twentieth of the norm; pathological excretion of phosphoethanolamine in the urine	Diminished alkaline serum phosphatase; excretion of phosphoethanolamine in the urine
Hypoproconvertinemia	Absence of Factor VII (proconvertin) which is required for normal coagulation; tendency to bleeding of skin and mucosa	Proconvertin reduced to 30–50% of the norm
Familial cretinism due to deficient de-iodization of diiodotyrosine	Myxedema, goiter; abnormal amounts of mono- and diiodotyrosine in blood after iodine administration	Diminished capacity of de-iodization of diiodotyrosine; abnormally high iodine-131 excretion in the urine after injection

TABLE 14—*Continued*

Anomaly	Homozygous State	Heterozygous State
Non-hemolytic congenital jaundice with kernicterus	Inability to couple glucuronic acid of bilirubin, salicylates, and hydrocortisone; bilirubin impregnation of the basal ganglia, opisthotonus, spasticity, rigidity	Diminished glucuronide formation after administration of salicylates
Parahemophilia (deficiency of Factor V [proaccelerin])	Severe bleeding with small injuries; tendency to bleeding of skin and nose	Factor V diminished to 40–60% of the norm; no clinical signs
Phenylketonuria	Deficiency of phenyl alanine hydroxylase, which is responsible for the transformation of phenyl alanine into tyrosine; phenyl alanine level in blood raised to 30 times the norm; pathological excretion of phenyl pyruvic acid, phenyl alanine, phenyl lactic acid, phenyl acetic acid, and acetyl phenyl glutamine in the urine; relative deficiency of pigment in hair and eyes; dry skin with tendency to eczema; idiocy up to imbecility	Raised phenyl alanine level in plasma (0.103 ± 0.016 mol/ml; controls: 0.068 ± 0.010) (from Knox and Messinger)
Stuart-Prower Factor deficiency	Stuart-Prower Factor (Factor X) below 10% of norm; tendency to bleeding especially with small injuries	Stuart-Prower Factor 20–70% of the norm; occasionally mild tendency to bleeding

disease, the different types of amaurotic idiocy, and Niemann-Pick's disease, no reliable methods for the determination of heterozygotes are yet known. In Niemann-Pick's disease and in juvenile amaurotic idiocy which must be separated from infantile amaurotic idiocy for clinical and genetic reasons, the lymphocytes of the peripheral blood contain conspicuous vacuoles. Such vacuoles have also been found with increased frequency in the heterozygotes of juvenile amaurotic idiocy, but this finding has not been confirmed by all investigators. In one family with four adults who had Niemann-Pick's disease, Pfändler found increased total lipid values in blood relatives who were considered to be heterozygous. However, the form described by Pfändler should probably be regarded as a genetically independent one, since two of four afflicted patients lived to the fifth decade, and one each to the fourth and third decade of life, which is quite unusual for Niemann-Pick's disease. In the usual Niemann-Pick's disease, the lipid values of serum are ordinarily normal even in homozygous patients. In xeroderma pigmentosum, the heterozygotes are conspicuous because of numerous, rather dark freckles which are heavy even on the arms. They are especially sensitive to sunlight. In the homozygous state, the sensitivity to light is considerably more pronounced. It may lead to photophobia, reddened conjunctiva, pigmented spots, telangiectases, atrophic skin regions, wartlike hyperkeratoses, superficial ulcers, and malignant deformation. Gradually, severe, scarred mutilation of nose, eyelids, ears, and hands sets in.

For further recessive disorders which are presumably due to enzyme deficiency, it is expected that future progress in the explanation of the pathogenesis will have the result that the diagnosis of heterozygotes will be possible. In the case of recessive hereditary diseases, the parents and children of patients are always certain to be heterozygotes, while 2/3 of the healthy siblings are heterozygotes. In any case, the biochemical investigation of hereditary diseases such as alkaptonuria, albinism, the different disorders of thyroid hormone synthesis, of oxalosis and others should, if possible, include the closest relatives.

The discovery of heterozygotes by means of refined tests is a beautiful example of how a hypothesis which was first based purely on statistical genetics—namely, the hypothesis of recessive inheritance—was confirmed quite independently of statistics, by clinical and chemical methods. The situation is clearest for those anomalies which are regularly recognizable in the heterozygous state, namely, in acatalasemia and the deficiency states of the

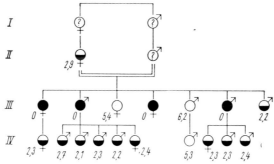

FIG. 22.—Pedigree with acatalasemia. For the sake of clarity, only those individuals are entered for whom blood catalase values were determined. The figures indicate the catalase units in the blood. Sharp separation between heterozygotes and the two homozygotes. (From Nishimura, Hamilton, Kobara, Takahara, Ogura, and Doi.)

coagulation Factors V, VII, and X. Several pedigrees of these anomalies show the following regularities in the recessive inheritance: Homozygous individuals always have two heterozygous parents and only heterozygous children —except for the extremely unlikely possibility that they are derived from one homozygous and one heterozygous parent, or that they themselves in turn have a heterozygous marriage partner. Homozygous patients can have homozygous, heterozygous, and homozygous normal siblings. Rare genes become homozygous relatively often through marriages between blood relatives (see Figs. 22–25).

Examples of a regularly recognizable effect of recessive genes in the heterozygous state are of didactic value. The question of why the heterozygous state can sometimes be definitely proved, but sometimes remains unrecognizable in the same disorder, is of theoretical interest. It should not be assumed that this is due to different alleles, perhaps in part stronger ones which therefore manifest themselves as heterozygotes and in part weaker ones which are concealed in the heterozygous state. For it is not infrequently found that, in one parent of a homozygote, the heterozygous gene has a distinct effect, such as the pathologically increased chloride and sodium values in perspiration in the case of cystic fibrosis of the pancreas, while the other parent cannot be distinguished in any way from a normal person. In the adrenogenital syndrome, increased elimination of pregnanetriol is found only occasionally in heterozygotes. How can this lack of homogeneity among heterozygotes as compared to the conspicuous homogeneity of the homozygous pathological and the relative homogeneity of the homozygous normal

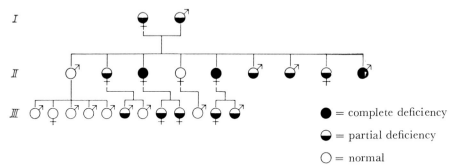

● = complete deficiency

◐ = partial deficiency

○ = normal

FIG. 23.—Factor V deficiency. Only examined patients are shown. (From Kingsley.)

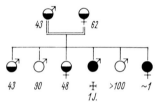

FIG. 24.—Factor VII deficiency. The figures give the Factor VII values in per cent. (From Hitzig and Zollinger.) For a boy who died early, no determination of Factor VII, which was then still unknown, was carried out but, according to the anamnesis, he presumably had a Factor VII deficiency (severe bleedings from mouth, nose, and intestine).

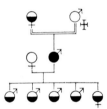

FIG. 25.—Stuart family with Factor X deficiency. I_2 had already died at the time of the study and could no longer be examined. Since he was a blood relative of I_1, and II_2 evidently was homozygous, I_2 must have been heterozygous. (From Graham, Barrow, and Hougie.)

control person be explained? The most obvious hypothesis seems to me to be that differences in the "normal" allele are expressed here. If, within the population, a gene locus is occupied by a series of different alleles with differing effects, the chance that a gene with an effect which is not yet pathological, but is "low-normal," will meet an equal gene is not great. In individuals who are heterozygous for a pathological gene of this series of alleles, however, the effect of such a low-normal allele becomes measurable. The situation here would be similar to that of sickle cell heterozygotes, who are hardly noticeable if they are simultaneously heterozygous for the gene for normal hemoglobin A, but who exhibit clearly pathological symptoms if they are heterozygous for another allele in the series, perhaps for hemoglobin C.

Speculatively, the parallel could be drawn even somewhat further. The great frequency of hemoglobinopathies is presumably based on the fact that the responsible genes confer a certain advantage in the heterozygous state, presumably in regard to resistance to malaria. In malaria regions, the population is polymorphic in the Hb_β series of alleles, that is, the locus is occupied not only by a single allele and very rare pathological exceptions, but, in notable portions of the population, by different alleles. Among recessive hereditary diseases in the European population, cystic fibrosis of the pancreas and the adrenogenital syndrome have the greatest frequency. Perhaps the heterozygous state confers an advantage here, also. But if one assumes such an advantage for heterozygotes, this presumably is not limited to the genes which are definitely pathological or even lethal in the homozygous state. It should rather be assumed that there are others among the numerous further possible alleles which determine a polymorphism that promotes survival. There is at present a lively discussion of borderline cases of the adrenogenital syndrome as well as of cystic fibrosis of the pancreas. Hirsutism, menstrual disorders, and mild hypertrophy of the clitoris are characteristic of the "borderline adrenogenital syndrome" (Gold and Frank). These are cases which, for clinical reasons, certainly cannot be included in the adrenogenital syndrome in the classical sense, and which yet exhibit striking parallels: high 17-ketosteroid excretion and high excretion of the pregnane complex, which decreases upon prednisone therapy. In the explanation of the problems involved in these borderline cases, which are perhaps considerably more frequent than is known at present, genetics is called upon to make an important contribution. Maybe "adult mucoviscidosis as a very frequent disease with dominant inheritance" (Bohn *et al.*) also belongs here. Certainly, the cases described under this title are not identical etiologically with the mucoviscidosis or cystic fibrosis of the pancreas of childhood. Perhaps, however, they are dependent on alleles of the same locus.

Attempt at a biochemical interpretation of the difference in effect between recessive and dominant genes

In contrast to the numerous examples of recessive enzyme deficiency states, there is no definite example of a dominant condition of enzyme deficiency. This leads to the conjecture that dominant hereditary diseases are due to the presence of a pathological gene product rather than to the lack of a normal gene product. An example of this can be found in hemoglobinopathies, which have been discussed under the concept of codominance. Dominant genes usually determine anomalies of tissue structure or of the shape of organs, without any demonstrable evidence of metabolic disorders (for examples, see Table 15). Such anomalies, which are more of a morphological nature, can be understood if one assumes that, in addition to a normal gene product, an abnormal gene product is fitted into the tissues as a building block. Of course, the primary gene product can be pictured here, also, as an enzyme which influences the synthesis of the building block only secondarily. In the structure of the cells and of the intercellular substance, a built-in pathological primary gene product as well as an abnormal enzyme product would lead to a permanent structural anomaly, even if the normal primary or secondary gene product were also present. In contrast, it is not so easy to visualize that an ineffective enzyme could do extensive functional damage to

TABLE 15

DOMINANT ANOMALIES OF TISSUE STRUCTURE AND SHAPE OF ORGANS

Anomaly	Clinical Features
Marfan syndrome	Anomaly of elastic tissue? Extreme asthenic habitus with spider fingers, funnel chest, scoliosis, lens luxation, insufficiency of aorta, aneurysms of aorta, median necrosis of the aorta
Ehlers-Danlos syndrome	Anomaly of collagen fibers, overextensibility of the joints; skin can be lifted far off, but snaps back elastically; slight tendency to injuries of skin, scars, calcifying hematomas
Osteogenesis imperfecta	Maturation disorder of the collagen fibers; thin bones, brittleness of bones, thin, blue, transparent sclera and tympanic membranes; overextensible joints; deafness due to otosclerosis-like changes and changes in the cochlea
Osler's telangiectasia hemorrhagica	Dilated small blood vessels with defective musculature and elastica; nosebleeding; arteriovenous fistulas in the lungs
Tuberous sclerosis (Adenoma sebaceum; M. Bourneville, M. Pringle)	Numerous hamartoma-like tissue malformations in brain (tuberous sclerosis), heart muscle (rhabdomyomas), skin (adenoma sebaceum), and kidneys (mixed growths)
Cystic kidneys of adults	Bilateral polycystic kidneys
Dysostosis cleidocranialis	Defects of the clavicles, late closure of the large fontanelle, pelvic fissure, stunted growth
Nail-patella syndrome (Turner-Kieser, onycharthrosis)	Defective development of the nails of fingers I to III; tendency to patellar and radius luxation; pelvic horns; tendency to "nephritis"
Typical cleft hand	Regularly cleft foot; hands not always involved
Chondrodystrophy	Dwarf growth; sunken root of nose; short limbs with normal trunk length; increased angle between lumbar portion of spine and sacrum
Multiple exostoses	Numerous exostoses, primarily of the proximal humeral epiphyses, the distal radial and ulnar epiphyses, the distal femoral and proximal tibial and fibular epiphyses; considerable deformation of the long bones; stunted growth

TABLE 16

Clinical Designation	Missing Enzyme and Its Function
Adrenogenital syndrome without high blood pressure	21-hydroxylase (conversion of 17-oxyprogesterone into 17-oxy-11-deoxycorticosterone)
Adrenogenital syndrome with high blood pressure	11-B-hydroxylase (conversion of 17-oxy-11-deoxycorticosterone into 17-oxy-corticosterone)
Acatalasemia	Catalase (conversion of hydrogen peroxide into water and oxygen; protection from H_2O_2-producing hemolytic streptococci)
Albinism	Tyrosinase (conversion of tyrosine into 3,4-dioxy-phenylalanine (DOPA) and of DOPA into DOPA-quinone, a precursor of melanin)
Alkaptonuria	Homogentisicase (conversion of homogentisic acid into maleyl-aceto-acetic acid)
Galactosemia	Galactose-1-phosphate-uridyl transferase (transfer of the phosphate from galactose-1-phosphate to glucose)
Glycogenosis, hepato-renal form (von Gierke)	Glucose-6-phosphatase (conversion of glucose-6-phosphate into glucose and phosphate)
Glycogenosis Type III according to Cori (abnormal glycogen)	Amylo-1, 6-glucosidase (conversion of glycogen into glucose-1-phosphate)
Glycogenosis Type IV according to Cori (abnormal glycogen, cirrhosis of liver)	Amylo-1,4–1,6-transglucosidase (conversion of glucose-1-phosphate into glycogen)
Hemolytic, non-spherocytic anemia	Phosphoglyceromutase: glycolysis (conversion of Nilsson ester into 2-phospho-glyceric acid)
Hypophosphatasia	Alkaline phosphatase (physiologic function unknown, involved in the mineralization of bone)
Hypothyreoses due to defective hormone synthesis; three independent forms known*	1.?? (conversion of iodide into elementary iodine) 2.?? (conversion of mono- and diiodotyrosine into triiodothyronine and thyroxin) 3. Iodotyrosine-dehalogenase (liberation of iodide from diiodotyrosine)
Non-hemolytic icterus of the newborn (with kernicterus)	? ? Enzyme contained in the microsomes of the liver (coupling of bilirubin and other substances such as menthol, salicylate, hydrocortisone metabolites to glucuronic acid)
Methemoglobinemia Type I	Diaphorase I. ? (reduction of methemoglobin to hemoglobin, together with reduced diphosphopyridine nucleotide)
Pentosuria	? ? TPN-Xylite (L-xylulose)-dehydrogenase (conversion of L-xylulose into xylite)

* Two further types are now definitely distinguishable.

TABLE 16—*Continued*

Clinical Designation	Missing Enzyme and Its Function
Phenylketonuria	? L-phenyl alanine hydroxylase (conversion of phenyl alanine into tyrosine)
Pseudocholinesterase deficiency	Pseudocholinesterase (physiological function unknown [see pp. 146 ff.])
Tyrosinosis	p-oxyphenyl pyruvic acid hydroxylase (conversion of p-oxyphenyl pyruvic acid into homogentisic acid)
Wilson's pseudosclerosis (hepatolenticular degeneration)	? Caeruloplasmin? Kinase in the liver, which forms, from a precursor (fraction C-D), C-C caeruloplasmin (physiological function unknown, involved in copper metabolism)

the metabolism as long as only the normal enzyme is also present. A good example of the differential nature of the primary defect in recessive and dominant modes of heredity is given by the hereditary methemoglobinemias. In the methemoglobinemias, a considerable part of the hemoglobin is present in its oxidized form, as methemoglobin. This leads to a compensating polycythemia. Clinically, cyanosis of the patient is the most prominent symptom. In recessive methemoglobinemia, an enzyme which brings about the reduction of methemoglobin by means of reduced pyridine nucleotides is lacking. Normally, methemoglobin is produced constantly in the metabolism but is immediately reduced to hemoglobin again through enzyme action. In recessive methemoglobinemia, the defect is in the removal of the methemoglobin which is produced. In the dominant methemoglobinemias, on the contrary, an abnormal hemoglobin is formed which is oxidized much more rapidly than ordinary hemoglobin. The changes in the dominant methemoglobinemias concern the globin portion of hemoglobin; so far, three different types are known, designated as Hb M_B, Hb M_S, and Hb M_M. In recessive methemoglobinemia, the defect in the reduction of methemoglobin can be counteracted temporarily by means of reducing substances such as methylene blue or ascorbic acid. In the dominant methemoglobinemias, reducing substances are ineffective.

The primarily structure-related effect of dominant pathological genes and the primarily metabolism-related effect of recessive pathological genes can be regarded only as a rough rule of thumb. There are also recessive hereditary anomalies of structure and form, such as the Ellis–van Creveld syndrome, in which one finds changes in the limbs similar to those in chondrodystrophy; also polydactyly, absence of tooth buds, and cardiac malformations, or the Bardet-Biedl syndrome with mental deficiency, obesity, polydactyly, and tapetoretinal degeneration; chondrodystrophia calcarea with irregularly calcified epiphyses, clouding of the lenses, and limited motility of the joints; pseudoxanthoma elasticum with crepe-rubber- or pigskin-like changes in the armpits and on the neck, vessel-like stripes in the back of the eyes, and tendency toward intestinal bleeding; Fanconi's panmyelopathy which is frequently associated with aplasia of the radius or thumbs; and many others. On the other hand, there are also several dominant anomalies which appear to be pure metabolic disturbances without primary morphological changes, such as pitressin-sensitive diabetes insipidus, renal diabetes mellitus, and essential hypercholesterolemia. In so-called constitutional liver dysfunction (icterus intermittens juvenilis Gilbert-Meulengracht), which is dominantly inherited, an enzyme seems to be missing which coupies bilirubin to glucuronic acid, so that the bilirubin cannot be eliminated in the normal way. In spherocytic hemolytic anemia, which is also dominantly hereditary, an enzyme-determined disorder of the glycolytic metabolism of the red corpuscles seems to be present. If one wants to make missing enzyme effects responsible for these dominant disorders, one could think of the possibility that pathologically altered, functionally incapable enzymes are formed which are like the normal ones to the extent that they enter into competition with them in the substrate and block the reaction. In general, it is best not to view the gene-determined lack of an enzyme simply as a purely quantitative problem. What is actually measured in enzyme determinations is not a chemical substance, but its activity. Perhaps, also, most recessive hereditary diseases are due less to a lack than to a pathological change in the enzymes which renders them ineffective. Such a possibility has been discussed for the rare cases of Wilson's hepatolenticular degeneration in which, exceptionally, the serum copper and the serum copper oxidase level were normal, although all other clinical and biochemical signs were present, such as tremor and disturbances of co-ordination, Kayser-Fleischer's ring of the cornea, increased copper excretion in the urine, hyperaminoaciduria, and slight disturbances of liver function. In such cases, perhaps the chemically determined caeruloplasmin is abnormal in its structure. Since the normal physiological role of caeruloplasmin in the organism is unknown, it is entirely within the realm of possibility that these cases with apparently normal caeruloplasmin are in reality cases with a modified enzyme.

According to more recent investigations, there are two different caeruloplasmin fractions which can be separated chromatographically and which are designated as fractions C–C and C–D. In hepatolenticular disease, only the normally predominant caeruloplasmin C–C is missing. Perhaps the primary defect does not involve caeruloplasmin, but an enzyme contained in the liver, which changes the C–D fraction into C–C. In heterozygotes, the C–C fraction is diminished.

Sex-linked heredity

DOMINANT X-CHROMOSOMAL HEREDITY

Up to this point, the discussion regarding dominant and recessive genes is valid for genes which are located in the twenty-two autosomes. For the inheritance of these genes the sex of the individual possessing them plays no role. However, there are other genes, whose hereditary transmission depends on the sex; we call these sex-linked, or better still, X-chromosomal genes.

We have seen that, in the female sex, two X chromosomes are found besides the twenty-two autosomes, while the male sex has one X and one Y chromosome, and that the sex chromosomes separate like a pair of alleles. If a dominant gene is located on the X chromosome, the consequences for heredity are extremely simple. Here, the dominant X-chromosomal gene provides merely a kind of marking of the X chromosome. With the aid of this marking, it is possible to follow the mode of inheritance of the X chromosome through the generations. An X-chromosomal dominant gene never passes from father to son, but always to all daughters. From the mother, however, the gene can be transmitted on the average to half the sons and half the daughters (Figs. 26, 27).

Only a few dominant X-chromosomal genes are known; but among these,

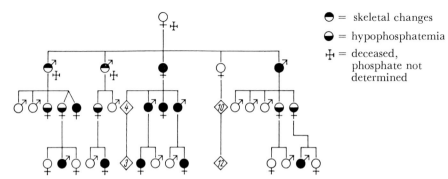

Fig. 26.—Excerpt from a kindred with X-chromosomal dominant hypophosphatemic rickets. It is characteristic for this mode of inheritance that all daughters of afflicted men are afflicted and all sons are free of the disease. (From Winters, Graham, Williams, McFalls, and Burnett.)

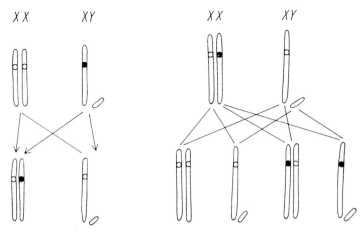

Fig. 27.—Chromosome scheme for Figure 26. Since homozygously normal women have only one kind of X chromosomes, the combination was drawn only for one maternal X chromosome.

there are two which are quite important in practice—vitamin-D–resistant rickets due to phosphate loss through the kidneys, and the deficiency in the erythrocytes of glucose-6-phosphate dehydrogenase, which can be the basis of severe hemolytic anemias after ingestion of vicia fava and of different drugs.

Vitamin-D–resistant hypophosphatemic rickets is distinguished from genuine vitamin-D–deficiency rickets by the fact that it occurs in spite of adequate vitamin-D prophylaxis, that the patients do not have muscular weakness, that hypocalcemia is absent, and that there is no increased excretion of amino acids in the urine. The experienced observer is often able to recognize such patients almost at first glance. They are small in stature, but strong at the same time, are distinctly bowlegged, more rarely knock-kneed, without other gross skeletal deviations, and often have distinctly elongated skulls. The low serum phosphate level is typical, usually amounting to less than 2 mg. per cent and ordinarily not becoming normal even upon vitamin-D therapy. The bone changes, on the other hand, can be cured to a great extent by administration of high doses of vitamin D. It is true that the treatment must at all times be carefully controlled, since there is a danger of overdosage. In adulthood, there are no instances of progressive osteomalacia. The patients are usually entirely fit for work. However, in later adult life, bone pains and increased bone deformations can again occur. Rickets-like deformations of the pelvis are not a part of the syndrome. Therefore, giving birth presents no difficulties. Women are usually less severely afflicted than men, in regard to both the organic phosphate level and bone changes. This is probably due to the fact that women have, in addition to the pathological X-chromosomal gene, a normal allele in the other X chromosome, while men are hemizygous for the pathological gene. Presumably, the short, bowlegged men less frequently have the opportunity of marrying. This is probably the reason why it took such a long time to detect the X-chromosomal mode of inheritance. Of course, the X-chromosomal dominant mode of inheritance cannot be recognized in the descendants of diseased women, but only in the descendants of diseased men.

The deficiency of glucose-6-phosphate dehydrogenase of the erythrocytes, the basis of favism, is also dominantly inherited through X chromosomes. For some persons, vicia fava, the broad bean, is a hemolytic poison. After ingestion of these beans, a severe feeling of illness results, with fever, vomiting, dizziness, rapidly increasing anemia, icterus, and hemoglobinuria. Favism is widespread primarily in Sardinia, Sicily, and Minorca. Its cause is an enzyme defect of the erythrocytes. The peculiarity of this enzyme defect is its surprisingly wide distribution. In Sardinia, 12 per cent of the population is afflicted; among Kurdish Jews, 30 per cent; among the Jews of southern Turkey, 8 per cent; and among Negroes in the United States of America, 10–14 per cent of the male population. This anomaly has been observed only in Sephardic Jews, never in Ashkenazic Jews. It has been found also among Greeks, Arabs, Persians, and Indians. For the physician, knowledge of this enzyme anomaly is of great significance, since it determines not only intolerance of broad beans but also pathological reactions to drugs. In this regard, primaquine, which is used for the treatment of malaria, is especially important, since frequently a considerable part of the population in the regions named must be given medication for malaria. Furthermore, hemolytic anemias may be triggered in individuals who lack glucose-6-phosphate dehydrogenase by sulfamide, naphthaline, synkavite, and furadantin.

If the homes of patients with glucose-6-phosphate dehydrogenase deficiency are entered on a map, it will be noted that, without exception, they come from regions where malaria is endemic. Is malaria then perhaps the cause? This cannot be assumed for the reason that, even generations after emigration to the malaria-free United States, the population groups of African and southern European origin retain the high frequency of the erythrocyte defect. The relationship is probably determined by selection. The malaria plasmodia as parasites of the erythrocytes are adapted in their metabolic requirements to the normal erythrocyte metabolism; for example, they utilize normal hemoglobin A. If the erythrocytes contain pathological hemoglobin types, perhaps sickle cell hemoglobin, the malaria plasmodia do not grow well. Sickle cell hemoglobin confers a certain degree of protection against malaria. Evidently, the glucose-6-phosphate dehydrogenase deficiency acts in a similar way. The malaria plasmodia do not need this enzyme itself. However, for their metabolism a high content of reduced glutathione is required. Glucose-6-phosphate dehydrogenase removes hydrogen from the glucose metabolism and transfers it to TPN; from TPNH, the hydrogen is transferred by glutathione reductase to the oxidized glutathione. (TPN is the abbreviation for the triphosphopyridine nucleotide; TPNH, that for its reduced form.) TPN is an important hydrogen transfer agent in metabolism. In glucose-6-phosphate dehydrogenase deficiency, the transfer agent of hydrogen for the reduced glutathione is thus missing. The malaria plasmodia find themselves in a state of metabolic deficiency. In this way, according to the plausible ideas of Motulsky, a state of enzyme deficiency confers an advantage in the defense against malaria. In regions with endemic malaria, the defect is bred preferentially.

From the genetic point of view, the glucose-6-phosphate dehydrogenase deficiency is especially interesting because we are able to compare here not only three possible genotypes—homozygously pathological, heterozygous, and homozygously normal—but five different ones, namely, in the male sex, hemizygously pathological and hemizygously normal and, in the female sex, homozygously pathological, heterozygous, and homozygously normal. No differences can be detected between homozygously normal and hemizygously normal genotypes, or between the homozygously pathological and hemizygously pathological. Differences can be shown, however, between heterozygously and hemizygously pathological cases, in that heterozygous women evidently are not afflicted as severely as homozygous women and as hemizygously pathological men. The pathological gene exerts a weakened effect beside its normal allele. In contrast to other enzyme defects, however, this effect cannot be due simply to a lack of the normal gene product, for hemizygously normal men also have only one normal allele and are nevertheless indistinguishable phenotypically from normal women in regard to the glucose-6-phosphate dehydrogenase content of the erythrocytes. Thus, the pathological gene must operate actively, that is, not only because of the absence of the normal gene product, just as we have postulated in other cases for dominant genes. This again does not fit the chemically detectable nature of the disorder, which seemed to be confined entirely within the framework of the conditions of enzyme deficiency. An explanation of this discrepancy, which is difficult to understand, between the positive effect of a gene and the phenic enzyme deficiency produced by it, was made possible through the investigations of Marks, Gross, and Hurwitz. The glucose-6-phosphate dehydrogenase content of the leukocytes and of the liver cells as well as of the young erythrocytes is normal in the patients. Evidently, the enzyme decreases with abnormal rapidity in the erythrocytes only as these age. The primary effect of the gene thus appears to have nothing to do with enzyme production, but only with the conservation of the enzyme content in the aging erythrocytes.

From the point of view of population genetics, the example of glucose-6-phosphate dehydrogenase deficiency is instructive, since even homozygous women are not too rare because of the high frequency of the gene. By means of the glutathione stability test, homozygous women or hemizygous men can be distinguished with certainty from the healthy persons used as controls. Approximately 15 per cent of the male Negroes and about 2 per cent of the female Negroes in the United States fall within the order of magnitude of the glutathione values which are found in hemizygous men. The gene frequency p for a hemizygous gene is equal to the frequency of the trait in the male population; the frequency for homozygous women is p^2. Expectation and observation are in satisfactory agreement ($p = 0.15$, $p^2 = 0.022$). Here, of course, the frequency of heterozygous women must be considerably higher, namely, $2pq$ ($p = 0.15$, $q = 0.85$, $2pq = 0.255$). It is true that methods developed up to now do not permit a reliable separation of heterozygous women from those who are homozygously abnormal, on the one hand, and from those who are homozygously normal, on the other. The fact that the glutathione curve for men shows separate distribution curves without overlapping, while that for women has a wide intermediate zone, in itself speaks very strongly in favor of an X-chromosomal condition with intermediate dominance.

Two additional dominant X-chromosomal anomalies likewise exhibit more pronounced disorders in the male sex—hypoplasia of the tooth enamel and keratosis follicularis spinulosa et decalvans. Hypoplasia of the tooth enamel, which has X-chromosomal dominance and which was studied in detail for the first time by Schulze, might in the male sex be called aplasia of the enamel. Roentgenologically, no enamel is recognizable, and histologically only a very thin, atypical layer of enamel is shown. In women, the enamel can be shown roentgenologically; it is rough, arranged in vertical ridges, and histologically of irregular structure (Fig. 28). In keratosis follicularis spinulosa et decalvans, the hair follicles of the eyebrows and lashes and of the cheek region are horny and bristle-like, and the skin in the nape of the neck is thickened and wrinkled. There is loss of hair in this region (baldness of the nape of the neck). After puberty, the horny bristles fall off and leave scars. In addition, one finds increased sweating and dystrophy of the cornea in this anomaly. Among female family members with the hereditary trait, the symptoms are considerably less pronounced. Here, eyebrows and lashes are not affected. The changes in the cornea can be detected only with a slit lamp; however, on the face and on the extended sides of the arms, the horny bristles are as pronounced as in the male sex. In the first published kindred having this anomaly, the mild manifestations in the female sex had not been

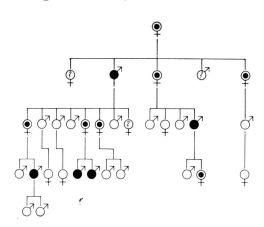

Fig. 28.—Kindred with X-chromosomal hypoplasia of the enamel, with intermediate dominance. Some of the healthy descendants of healthy family members have been omitted for the sake of clarity. (From Schulze.)

noted, so that the published pedigree showed the picture of recessive X-chromosomal inheritance. This also is an example of the fact that the difference between dominant and recessive can be quite relative.

Women are afflicted twice as frequently as men with dominant X-chromosomal anomalies, since there is only one X chromosome in the male for every two X chromosomes in the female. In the two pedigrees of Figures 26 and 28, there is not an exact 2:1 ratio. But this is due to the fact that it was necessary to select pedigrees with afflicted men who had children, for the purpose of demonstrating the peculiarities of the mode of inheritance.

RECESSIVE X-CHROMOSOMAL OR SEX-LINKED HEREDITY

Recessive X-chromosomal genes are considerably more frequent than dominant ones. Perhaps the frequency of X-chromosomal dominant inheritance is somewhat underestimated, since the typical picture of recessive sex-linked inheritance can be read at first glance from a pedigree, while dominant X-chromosomal inheritance is recognizable only through careful observation, and only under favorable circumstances. Presumably, the anomalies described as dominant include several which are X-chromosomal, for example, a type of microphthalmia and a special type of ectodermal dysplasia with missing rudiments of the incisors of the lower jaw and of several additional teeth, low-degree hypotrichosis, and normal sweat secretion (Figs. 29–31).

Important anomalies such as red-green color blindness, hemophilia A and B, the most frequent type of progressive muscular dystrophy, and nephrogenic diabetes insipidus are recessively X-chromosomal. The classical exam-

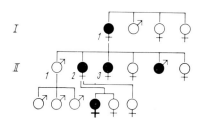

FIG. 29.—Hidrotic ectodermal dysplasia. Presumably dominant X-chromosomal inheritance.

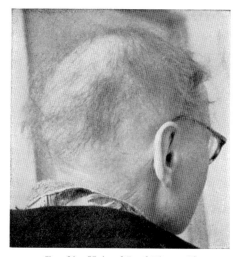

FIG. 30.—Hair of I₁ of Figure 29

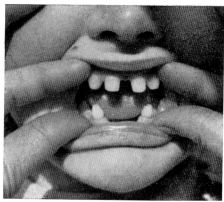

FIG. 31.—Teeth of II₃ of Figure 29

ple of recessive X-chromosomal inheritance is hemophilia (Fig. 32). Its mode of inheritance follows the X chromosome through the generations. Phenotypically, however, the gene is apparent only in the hemizygous state in the male. In women, the gene, which is located on one of the two X chromosomes, is masked by the allele on the second X chromosome. A woman who carries the gene (female carrier) transmits it on the average to half of her daughters, who now become carriers themselves, and to half of her sons, who manifest the disease. Transmission from father to son is not possible, since the father's X chromosome is passed on only to the daughters, and to all daughters. Therefore, all children of a male hemophiliac are phenically healthy, but all daughters are carriers of the gene. The sons of a male hemophiliac are not "hereditarily burdened." Their descendants are likewise free of the trait. It is characteristic for a pedigree with a recessive X-chromosomal hereditary disease that cases of the same kind are found in the brothers of the mother and of the maternal grandmother. In considering the frequency of recessive X-chromosomal hereditary diseases, in each family anamnesis questions about the brothers of the mother and of the maternal grandmother as well as about sons of aunts on the maternal side are especially important. In the counseling of families with hemophilia, the investigations of Inga Nilsson, Blombäck, Thilén, and Francken have brought about some very

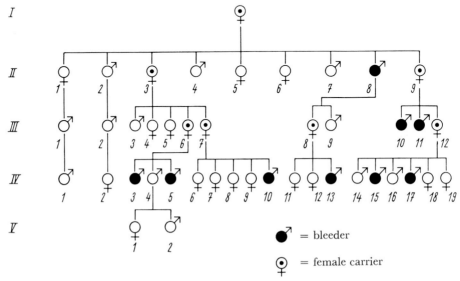

FIG. 32.—Bleeders among the descendants of Queen Victoria of England (I_1). II_8: Leopold. III_7: Alice, married Nicholas II, tsar of Russia. III_{12}: Victoria Eugenia, married Alfonso XIII of Spain. IV_3: Waldemar of Prussia. IV_5: Heinrich Sigismund. IV_{10}: Alexis, tsarevich of Russia. IV_{13}: Rupprecht. II_2 (Edward VII of England) and III_1 (Wilhelm II of Hohenzollern) had no "hereditary burden." (From Marx.)

important advances. Among the sisters and maternal aunts of hemophiliacs, the carriers of the gene can now be distinguished with great accuracy from those women who do not possess the pathogenic gene. One can now tell the latter that they can have children without worry. The gene for hemophilia is not absolutely recessive. The coagulation defect of the plasma of patients with hemophilia A can be corrected decidedly less effectively by plasma from female gene carriers of hemophilia A than by plasma of normal persons. The carriers of the gene occasionally also exhibit an increased tendency toward bleeding, especially after tooth extractions and childbirth.

Clinically and from the point of view of hereditary biology, classical hemophilia seemed to be a homogeneous disease. Differences in the severity of the disease were known. In some cases bleeding was relatively mild, and in others all afflicted members suffered early from massive bleeding of the joints and severe disablement. It is to be assumed that such different expressions of hemophilia are due to allelic mutants of the same gene. It has been possible to prove with the newer quantitative methods for the determination of anti-hemophilic globulin that the afflicted members of a family agree quite closely in the quantity of antihemophilic globulin.

It has also been known for several years that, beyond these quantitative differences, classical X-chromosomal hemophilia is not a homogeneous disease but that it must be subdivided into two qualitatively different types. This realization is not only of academic interest but also of significance for therapy, since the treatment of severe bleeding by blood transfusions has a considerably longer-lasting effect in one kind of hemophilia (B) than in the other (A). The recognition of this non-uniformity of hemophilia resulted first from the observation that it was possible in some cases to achieve a normal coagulation capacity if the plasma from two different patients with hemophilia was mixed, although each individual plasma showed a pronounced delay in coagulation. Such a reciprocal correction of the coagulation disorder, however, could never be achieved with two plasma samples from members of the same family. It was therefore obvious to suspect that two different forms of hemophilia existed, and that each of them lacked a different substance necessary for normal coagulation. In the meantime, it has been possible to determine more closely the differing characteristics of the two antihemophilic globulins. In hemophilia A, antihemophilic globulin A (Factor VIII) is missing; this normally occurs in plasma but, after coagulation, disappears from the remaining serum. Antihemophilic globulin B, on the other hand, which is missing in hemophilia B, can still be found in the serum of normal persons; that is, it is not used up by coagulation. A statistical analysis of about six hundred cases of hemophilia in the literature gives a frequency ratio of hemophilia A to hemophilia B of about 5:1. Hemophilia B, also, occurs in forms of differing severity. Within the same family, however, the patients are afflicted with equal severity.

Hemophilia is a relatively rare disease: there is one hemophiliac for approximately every 25,000 male persons. The gene frequency in the male sex corresponds, for a recessive X-chromosomal disease, to the frequency of the anomaly in the male sex—0.00004. If it is assumed that chance combination of genes is prevalent in the population and that hemophilic patients marry as frequently and have children as often as normal men, the calculated theoretical frequency of homozygous women would be 0.0000000016 ($p = 0.00004$, $p^2 = 0.0000000016 = 16$ per 10 billion). In reality, however, the theoretical assumptions of the calculation are not valid, for, if the hemophilic patients have no children, or fewer than other persons, the figures are altered. If hemophilic patients never reproduced, the result would be that a third of all cases were due to new mutations (see below, pp. 114 ff.). Thus, only 2/3 would have inherited their pathological gene from a heterozygous mother. The heterozygote frequency here would turn out to be $2/3 \times 2p$

$= 0.000053$, instead of $2p = 0.00008$ as in the case of normal propagation by the patients. The actual value for the frequency of heterozygous women should lie between these two values, and closer to 0.000053, since the reproduction rate of hemophiliacs is low. If we now wish to correct the calculation of the frequency of homozygous women, we must insert the gene frequency in the female sex, which equals half the heterozygote frequency and therefore lies between 0.000026 and 0.00004, and multiply it by the gene frequency in the male sex, which, however, must be reduced to a value corresponding to the effective fertility of hemophilic patients. The effective fertility of the hemophiliac amounts to about 30 per cent of the average fertility. Thus, the gene frequency of 0.000026 to 0.00004 in the female sex would have to be multiplied by the reduced gene frequency in the male sex of $0.00004 \times 0.3 = 0.000012$, in order to obtain a frequency of homozygous women which corresponds somewhat better to the facts. We obtain 3.1 to 4.7×10^{-10}. However, this figure neglects the cases of homozygous women who are produced by a combination of a new mutant with a gene inherited from the parents,

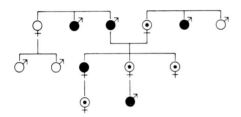

FIG. 33.—Pedigree containing a homozygous woman patient with hemophilia. (From Israels, Lempert, and Gilbertson.)

or by the combination of two new mutants. But this does not cause any appreciable change in the order of magnitude. Much more significant is the deviation from the theoretical assumptions of the calculation, which results because the genes do not mix at random in the population but, within certain "isolates," equal meets equal much more frequently than would be expected in a random mixture. The basic consideration remains, however, that the probability of occurrence of homozygous women is extraordinarily small for rare recessive X-chromosomal genes. The first definite case of hemophilia in a woman was published in 1951. The mother of the patient was a carrier, as could be recognized from the fact that she had a hemophilic brother, and the father of the hemophilic woman patient was himself a bleeder (Fig. 33). In another family, also, a girl with typical hemophilia A had been observed (Fig. 34). The mode of inheritance otherwise showed the typical picture: four male patients were related to one another through healthy women carriers. The father of the girl, however, was not a bleeder, as should have been expected with a homozygous woman patient. The solution of the puzzle came about through chromosome studies in cell cultures. The supposed girl had the chromosome set of a male individual. She was not homozygous, but hemizygous. An analogous observation was made also in a family with recessive X-chromosomal progressive muscular dystrophy. Here again, a female patient who was afflicted, contrary to the rule, was genetically male. In rare recessive X-chromosomal anomalies which occur in exceptional cases in women, one must always consider this possibility of a disorder of sex differentiation. These rare cases hardly have practical significance, but they are an impressive example of the precision with which the genetic theory agrees with the facts, if one does not let oneself be deceived by misleading appearances.

Disturbances of red-green color vision which are due to X-chromosomal genes are of considerable theoretical and practical interest. They are summarized under the not completely appropriate designation of "red-green color blindness" (Daltonism). Different degrees of red color blindness from protanomaly to extreme protanopia and of green color blindness from deuteranomaly to extreme deuteranopia, are distinguished. In abbreviated form, one also speaks of protan and deutan individuals, meaning collectively the

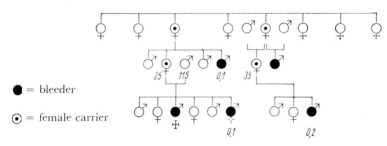

FIG. 34.—Family with classical hemophilia A. In the third generation, manifestation in a child with external appearance of a girl, with somewhat small labia minora, somewhat large clitoris, and uterus not palpable rectally, in whose skin and leukocytes no female sex chromatin could be detected, and in whom the sex chromosome constitution XY was found. The figures give the values of antihemophilic globulin in per cent of the norm. Some healthy descendants of healthy family members are omitted. (From Nilsson, Bergman, Reitalu, and Waldenström.)

persons having the different degrees of the two disorders. The total frequency of red-green color blindness in the male sex amounts to about 8 per cent; of this, protan individuals constitute approximately one-fourth and deutan individuals, approximately three-quarters. In the hemizygous male, only a single X chromosome is present; therefore, the traits of red-green color blindness cannot be masked. The frequency of the anomaly in the male sex, therefore, corresponds directly to the gene frequency in the population. If red-green color blindness were a homogeneous anomaly, the frequency of homozygous and therefore equally red-green color-blind women could be calculated simply from the gene frequency $p = 0.08$. The probability that both X chromosomes of the woman carry the responsible gene would be $0.08 \times 0.08 = 0.0064$. Comparative tests of color sense in both sexes have shown, however, that red-green color blindness occurs less frequently in women, as is shown in Table 17.

The difference between the value of 0.0064 expected theoretically and the value of 0.004 to 0.005 which is actually observed is based on the fact that red-green color blindness is not homogeneous genetically. A gene for a protan disorder which coincides in a woman with a gene for a deutan disorder does not lead to a combined disorder of the color sense, but to no disturbance of color vision at all. Only homozygous deutan women and homozygous protan women are incapable of color vision. On the other hand, doubly heterozygous deutan-protan women usually can distinguish red and green without error on the anomaloscope. The calculation of the frequency of color-sense disorders in women thus must be broken down into two parts, and we obtain, for a gene frequency for deutan genes of 0.06 and of protan genes of 0.02, the corresponding homozygote frequencies of $0.06 \times 0.06 = 0.0036$, and of $0.02 \times 0.02 = 0.0004$. By addition, these two frequencies give a total of 0.004 homozygous and therefore color-defective women, which corresponds approximately to the empirical figures.

The high frequency of red-green color blindness has made it possible to obtain information about the genetic sex of intersexual individuals in cases of disorders of embryonal sex differentiation which cannot be assigned simply to the male or female sex. Thus Sørensen has proved that red-green color blindness occurs as frequently in patients with hypospadias as among normal

TABLE 17

FREQUENCY OF RED-GREEN COLOR BLINDNESS

Country	Men	Women
Norway..............	0.08	0.0044
Switzerland..........	0.08	0.0043
France..............	0.09	0.0050
Scotland............	0.08	0.0056

men. He was thus able to show for the first time that the old concept is false, according to which hypospadias patients are supposed to be genetically female individuals with partial sex reversal.

Refined methods of investigation frequently make it possible even in the case of recessive X-chromosomal anomalies to recognize mild symptoms in heterozygous women. Hemophilia has already been mentioned in this connection. With hemophilia B, heterozygous women appear to tend toward bleeding even more frequently than with hemophilia A. The observation is to be noted that, especially in families with relatively mild hemophilia A, in whom the antihemophilic globulin A is not diminished to a very high degree, heterozygous women tend toward bleeding especially frequently. If one explains the pathological gene effect simply by the lack of antihemophilic globulin, this effect of the heterozygous genes cannot be understood. In addition to the pathological gene, a heterozygous woman has a normal allele, and under the assumption of the purely quantitative disorder of the gene product, it is difficult to understand why a heterozygous woman does not produce just as much gene product, in this case antihemophilic globulin, as a normal man who also has only one normal allele. Therefore, the possibility was thought of that the pathological gene leads not to a lesser production of antihemophilic globulin but to the production of a degenerate, functionally inferior gene product which resembles the effective gene product chemically to such an extent that it can enter into the reaction and block it. More precise chemical isolation and characterization of the antihemophilic globulin is needed to make this concept more definite. In general, however, much can be said in its favor. It is more readily reconcilable with our concepts of the model of primary gene action, that pathogenic genes lead to pathological gene products, than that a pathogenic gene produces no gene product at all. However, since we are unable to detect most of the gene products, such as the enzymes or the coagulation factors, by chemical means, but measure them only by their effects, for the present the question whether an abnormal gene product or even no gene product at all is present cannot be answered. All examples of heterozygous manifestations of X-chromosomal genes speak in favor of degenerate gene products.[1] Table 18 lists manifestations of different X-chromosomal genes in the heterozygous state; these can be designated as recessive in practical clinical terminology.

The designation "sex-linked" usually leads the layman to the erroneous concept that "sex-linked" anomalies must be tied to a specific sex. Therefore, it is preferable to speak of X-chromosomal anomalies. Accompanying the unclear concept of anomalies linked with the male sex, one often finds the idea that there must also be anomalies which are linked with the female sex. However, owing to the mechanism of chromosomal sex determination, this is not possible, since there is no specifically female chromosome. Nevertheless, it would be conceivable that some rare anomalies occur only in female individuals, because of a special genetic mechanism. Such a phenomenon has not been demonstrated in man, but it is known in Drosophila flies. If a recessive X-chromosomal gene in the hemizygous state leads to such a severe developmental disorder that no viable product can develop, but if, in the heterozygous state, the effect is weakened to such an extent by the normal allele that viability is possible, then such an anomaly will be transmitted from the mother to half of her daughters, while all of the sons who can be observed are healthy. The hemizygous carriers of the gene who died as embryos are not subjected to observation. In Drosophila, the notch-wing anomaly follows this type of inheritance. Here the afflicted female animals have wings with notched edges. The male zygotes with the responsible gene die off very early. In man, one must count on the possibility that a rare skin disease which occurs almost exclusively in the female sex, and which has been observed a number of times to be inherited by the daughter from the mother, follows such a mode of inheritance. This disease is the so-called incontinentia pigmenti Bloch-Sulzberger. In this anomaly, starting at birth, red papules, blisters with eosinophilic cells and hyperkeratoses occur predominantly on the trunk of the body and on the limbs, healing toward the end of the first year of life and leaving behind black-brown pigmented spots in the shape of stripes and whirls. Not infrequently, there is also an involvement of the nervous system, with spastic paralyses and epileptic seizures, as well as of the eyes with pseudoglioma. Almost always several permanent teeth are missing (Figs. 35 and 36).

[1] Recently, Mary Lyon has advanced the ingenious hypothesis that in each cell of a female individual only one of the two X chromosomes is active and that this is not the same X chromosome for all cells. Thus, women heterozygous for X-chromosomal genes would be functional mosaics. This is probably a better explanation of heterozygote manifestation.

TABLE 18

RECESSIVE X-CHROMOSOMAL ANOMALIES WITH PARTIAL MANIFESTATION IN
HETEROZYGOUS FEMALE CARRIERS OF THE GENE

Anomaly	Hemizygous and (as far as known) Homozygous State	Heterozygous Partial Manifestation
Angiokeratoma corporis diffusum Fabry	Scattered small aneurysms of the vessels in the skin with hyperkeratoses; albuminuria, progredient kidney insufficiency; clouding of cornea; phosphatide storage in different organs	Skin inconspicuous; typical kidney lesions with foam cells
Anhidrotic ectodermal dysplasia	Absence of numerous teeth (oligodontia); thin, sparse hair; secretion of perspiration absent, therefore deficient temperature regulation; hyperthermias	Hypodontia (absence of several teeth); normal secretion of perspiration
Albinism of the eye	Grey-blue, transparent iris, absence of pigment in back of the eye, visibility of the chorioideal vessels, hypoplasia of the macula, nystagmus, nodding of head	Pigment deficiency of back of the eye
Chorioideremia	Progressive atrophy of the chorioidea and retina; whitish fundus with islands of chorioidea and extremely thin retinal vessels; progredient visual disorder extending to blindness	Scattered or grouped granules of pigment, especially in the posterior pole, decreasing toward the periphery; no visual disorders
(Nephrogenic) diabetes insipidus	Specific weight of urine usually below 1005, always below 1010; no influence of pitressin; with insufficient administration of fluids, inhibition of growth and feeble-mindedness; in infancy and early childhood, life endangered by dehydration; fever with thirst; thickening of stools	Specific weight of urine, even in thirst experiments, hardly ever above 1020; occasionally polydipsia
(Diencephalic-pituitary) diabetes insipidus (X-chromosomal type; there is also a clinically indistinguishable autosomal dominant type)	Concentration of urine between 1001 and 1005, can be influenced by pitressin; polyuria	Sometimes slight limitation of concentrating capacity; sometimes, in the last months of pregnancy, manifest diabetes insipidus
Hemophilia A	High degree of decrease of antihemophilic globulin A; severe coagulation disorder	Especially in the milder forms, decrease of antihemophilic globulin; slight tendency to bleeding
Hemophilia B	Absence of antihemophilic globulin B; severe coagulation disorder	Tendency to bleeding of skin; sometimes diminished antihemophilic globulin B
Ophthalmoplegia externa with myopia	Paralysis of external eye muscles; Achilles and patellar tendon reflexes absent	Achilles and patellar tendon reflexes absent; no eye muscle paralysis
Red-green color blindness	Protan disturbances (red defect) and deutan disturbances (green defect)	Slight defects in red-green vision which are not demonstrable with the Ishihara Tables can be demonstrated in about 50% of the heterozygotes with the anomaloscope

PARTIAL SEX LINKAGE

On the basis of chromosome studies on the rat and other animals, it has been suspected that the X chromosome and the Y chromosome are in part homologous. Genes in this hypothetical common segment of the two sex chromosomes would have to show certain peculiarities in regard to inheritance. If, to begin with, we disregard crossing over, the situation becomes rather simple. A dominant, partially sex-linked gene would show, depending on whether it is located on the X chromosome or the Y chromosome, the usual inheritance of dominant X-chromosomal genes or a "holandric" inheritance, that is, hereditary trans-

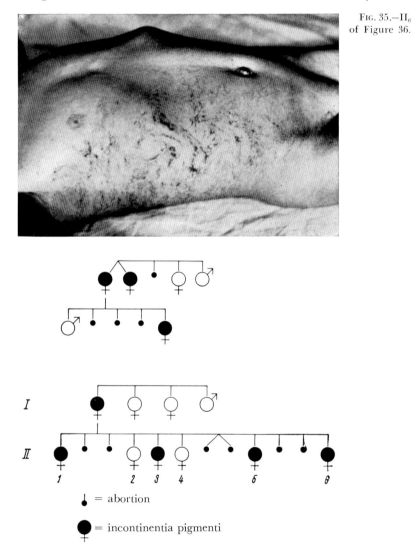

FIG. 35.—II$_6$ of Figure 36.

$\dagger$ = abortion

$\bullet$ = incontinentia pigmenti
$\dagger$

FIG. 36.—Families with incontinentia pigmenti Bloch-Sulzberger. Example of a hemizygous, prenatally lethal, heterozygously pathological gene (?).

mission from afflicted men to all their sons. Recessive, partially sex-linked genes, on the other hand, would be distinguished by the fact that the afflicted siblings would always be of the same sex—boys only, if the gene inherited from the father was located on the homologous section of the Y chromosome; and girls only, if it lay on the homologous section of the paternal X chromosome. Because of crossing over, of this rigid rule only certain frequency relationships would be left. Partially sex-linked recessive hereditary diseases would occur more frequently among siblings of the same sex than would be expected if there were no dependence on sex. Some diseases thought to be examples of incompletely sex-linked genes are total color blindness, xeroderma pigmentosum, one type of epidermolysis bullosa, Oguchi's disease, and one type of retinitis pigmentosa. It has even been attempted to calculate numerical values for the frequency of crossing over, using the frequency of deviations from the exclusive affliction of one sex within sibships. On this basis, a chromosome chart of the homologous section of the X and Y chromosomes has been constructed. It is known from Drosophila research that the frequency of crossing over between two linked genes depends on the distance between them in the chromosome. In partially sex-linked genes, one does not determine the crossing-over frequency between two genes. Instead, one determines the crossing-over frequency between the partially sex-linked gene and the non-homologous section of the sex chromosomes which is responsible for sex determination and which behaves in this regard like a simple Mendelian gene. On the hypothetical chromosome chart of the homologous section of the sex chromosomes, the genes for total color blindness, xeroderma pigmentosum, epidermolysis bullosa, etc., were entered at different distances from the point of transition from the homologous to the non-homologous section. The theory of incompletely sex-linked genes is on weak ground in regard to its morphological foundation as well as in regard to statistical genetics. We need not be further concerned with it here.

Y-CHROMOSOMAL GENES?

If a gene lies on the Y chromosome, it must be transmitted from the father to all sons, and from the sons on to all male descendants. In the older literature, several examples of such a "holandric" mode of inheritance were published, which also have found their way into many textbooks. None of these examples is confirmed (see above, p. 10).[2]

Series of allelic genes (multiple allelism)

The concept of multiple allelism should not be regarded as an extension of the discussion up to this point, but rather as its obvious consequence which has already been implicitly assumed. Each individual has only two identical or non-identical genes at each gene locus. However, in the population there are presumably not only two but several, or many, alleles for most gene loci. For the gene loci which are responsible for the antigenic properties of the red blood corpuscles, we usually know several alleles. The best known is the ABO blood group system, within which there are not only allelic genes for

2 Recently, better evidence of Y-chromosomal inheritance of hairy ear rims has been produced (Gates and Bhaduri, 1961), but the subject remains controversial (Sarkar, Banerjee, Bhattacharjee, and Stern, 1961).

A, B, and O, but different sub-groups A_1, A_2, etc. The genes which are responsible for hemoglobin A, sickle cell hemoglobin, and hemoglobin C are also multiple alleles. Presumably still other pathological hemoglobin types belong to the same series. However, the number of families observed in which several hemoglobin types occurred simultaneously has been too small to provide a clear picture of the genetic conditions. The genes of the protan and deutan series also are multiple alleles. It has not been decided definitely whether both series belong to a single series of alleles. Some authors are of the opinion that two allele series are involved (two-locus theory of red-green color blindness). A decision will become possible only after extensive studies of linkage (see p. 100) have been made.

Pleiotropism and polyphenia. One gene
determines several characteristics

So far, we have simply assumed the pleiotropic or polyphenic effect of genes without a special explanation. This effect means that a certain gene determines not only a single characteristic, but a series of different characteristics which often do not show any recognizable relationship to one another. There is a recessive X-chromosomal gene which causes eczema, thrombopenia, and highly diminished resistance to infection, although, up to now, no internal relationship between these phenomena has been discovered. It is equally unclear why another intermediately dominant X-chromosomal (perhaps autosomal with sex-dependent action) gene simultaneously causes progressive pyelonephritis with hematuria and deafness of the inner ear, or why the X-chromosomal hereditary Lowe syndrome is accompanied by a metabolic disorder with excretion of organic acids in the urine, rickets-like skeletal changes, clouding of the lenses, and glaucoma. All of these diseases pose interesting pathogenic or, seen from the genetic point of view, phenogenetic problems. In other diseases, the mechanism which interrelates the different phenomena and makes it possible to trace them back to a single cause is adequately explained. As examples of this, we can cite the adrenogenital syndrome and galactosemia.

In the adrenogenital syndrome, the production of cortisol (17-hydroxycorticosterone) from 17-hydroxyprogesterone is upset. Cortisol is critical for the self-regulation of adrenal cortex function. A normal cortisol level in the blood slows down the hypophysis, so that it does not release an excessive amount of adrenocorticotropic hormone. The adrenocorticotropic hormone has the task of stimulating the growth of the adrenal cortex and the production of its hormones. In cortisol deficiency, the adrenocorticotropic hormone (ACTH) is released by the hypophysis in increased amounts. This hormone leads to increased steroid production in the adrenal cortex, but only up to the stage of 17-hydroxyprogesterone. The increased androgenic hormones perhaps originate partly in the 17-hydroxyprogesterone, but in part they are probably also produced directly under the stimulating influence of the

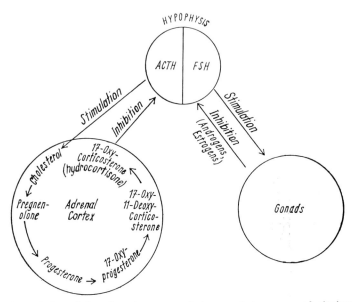

FIG. 37.—Self-regulation of the adrenal cortex and the gonads by means of pituitary hormones. Normal conditions. 17-Oxy-corticosterone = 17-hydroxycorticosterone.

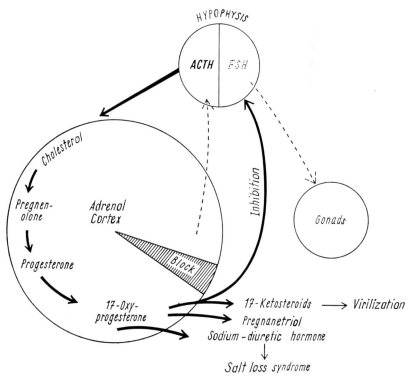

FIG. 38.—Blocking of corticoid synthesis in the adrenogenital syndrome, and secondary consequences in endocrine self-regulation. 17-Oxyprogesterone = 17-hydroxyprogesterone.

ACTH in the adrenal cortex. The clinical signs of the adrenogenital syndrome can be divided into three groups:

1. Effects of diminished cortisol production. The increased ACTH production and also presumably low resistance to infections and to surgical shock are included here.
2. Effect of ACTH. Hyperplasia of the adrenal gland and increased hormone production are included here.
3. Effects of increased androgen production, which are by far the most prominent in later childhood and adult life, while the salt-loss syndrome with vomiting and collapse can be the primary threat in infancy. The androgen effect in girls leads to masculinization of the genitals up to a pseudohermaphrodite or even pseudomasculine type, and in boys, to pseudopubertas praecox, i.e., to a premature development of the secondary sex characteristics without corresponding maturation of the testicles. The androgenic and likewise the increasingly produced estrogenic hormones, in their turn, inhibit the gonadotropic function of the hypophysis. An insufficient amount of follicle-stimulating pituitary hormone is formed, so that the onset of ovarian function is not regulated properly and maturation of the testicles fails to occur. The pathogenesis of the salt-loss syndrome has not been completely explained. There is some evidence favoring the theory that it is produced by increased formation of a sodium-diuretic hormone, and that it thus also is a part of the consequences of overproduction of adrenocortical steroids.

Figures 37 and 38 are intended to give a schematic summary of the normal relationships and their disturbance in the case of the adrenogenital syndrome.

It is easy to influence the adrenogenital syndrome therapeutically by adrenocortical hormones having a cortisol-like effect. When these hormones are administered, the ACTH level in the blood is lowered, the 17-ketosteroid secretion in the urine diminishes quickly to normal values, and the masculinization phenomena disappear to the extent that they are capable of retrogression. But therapy must be started early and continued throughout the entire lifetime.

It is a question of definition whether the adrenogenital syndrome is to be designated as an example of polyphenic gene action. The primary gene effect is of a homogeneous nature; the composite picture comes about only because of secondary regulating mechanisms and "polyphenic" hormone effects. If we had not been able to explain the pathogenesis of the disorder so well, and if we knew nothing of the nature of hormonal effects and regulatory mechanisms, we would not hesitate for an instant to regard the adrenogenital syndrome as a model specimen of a polyphenic gene effect. Presumably, however, the difference between the primarily homogeneous nature with secondary polyphenia, and the apparently primary polyphenia of other hereditary diseases is not a difference in principle but only in the extent of our knowledge. In most other hereditary diseases with polyphenic symptoms, we do not yet know the internal relationship, but there is good reason to assume that such an internal relationship always exists and that, in the end, a well-defined basic defect is the cause. To explain the pathogenesis a search for the basic defect is of decisive significance.

A recessive anomaly whose diverse symptoms, which at first glance can hardly be combined under a common denominator, are expressions of a specific enzyme deficiency, is galactosemia. In galactosemia, the primary defect consists of a deficiency of galactose-1-phosphate-uridyl transferase. This brings about an accumulation of pathological amounts of galactose-1-phosphate in various tissues, which finally are damaged by the toxic concentration, so that cirrhosis of the liver results, the glomeruli allow protein to pass through, and the kidney tubules can no longer sufficiently reabsorb the amino acids; therefore, aminoaciduria results. Clouding of the lenses occurs, similar to the way in which one can produce it in rats by giving a diet rich in galactose. The decrease in glucose in the blood, which can lead to cramps and feeble-mindedness, perhaps is due to damage of the liver which can no longer degrade its glycogen in the normal way. All phenomena of galactosemia can, to a great extent, be prevented and partially cured if the children are given a diet which does not contain any lactose. Administration of the enzyme is not possible in this case. Even if the enzyme were available in sufficient quantity, it still would not be possible to apply it in the location of its effectiveness, namely, in the interior of the cells. The practically complete normalization of the entire severe state of illness by a galactose-free diet is a clear indication that all the diverse manifestations in the organs are indeed due to the toxic effect of galactose, which cannot be further metabolized in the cells.

Polyphenic gene effects have repeatedly evoked speculations, which change with the current fashion. In the early years of endocrinology, it was attempted to trace back all kinds of genetically determined syndromes to hormone effects. It is a remnant from this period that the Bardet-Biedl syndrome is still discussed in some textbooks among the endocrine diseases even though hypogonadism is not a regular and clinically significant phenomenon in this case and recedes into the background compared to the other symptoms such as obesity, feeble-mindedness, and retinitis pigmentosa. With the increasing perfection of endocrinological methods of investigation, the untenability of the earlier speculations about endocrine causes became evident in numerous hereditary diseases.

Speculative interpretations were now confined more to the midbrain. The primary gene effects were still visualized as predominantly tied to organs, corresponding to the determinants of the germ plasma theory of Weismann, and a control center regulating the multiplicity of symptoms was looked for. The midbrain offered such a solution. It has been impossible, however, to advance plausible arguments for a gene effect located primarily in the midbrain, such as was postulated for quite a number of syndromes.

The most persistently maintained concept has been that gene effects are limited selectively to one of the three germinal layers—ectoderm, mesoderm, or endoderm—and that some disorders could best be understood in this manner. As a matter of fact, genes do not show much respect for the boundaries of the germinal layers, even though there are individual gene effects which manifest themselves primarily in the derivatives of one germinal layer. The latter probably depends more on the specific chemical nature of the gene product, and on the role this product plays in the derivatives of a germinal layer, than on the localization of the gene effect in a specific germinal layer. The basic defect of the Marfan syndrome shows itself primarily in derivatives of the mesenchyme; the same is true for the basic defect of osteopsathyrosis. Nevertheless, the two diseases are independent, well-distinguished syndromes, which can hardly be confused with each other and which have only a few

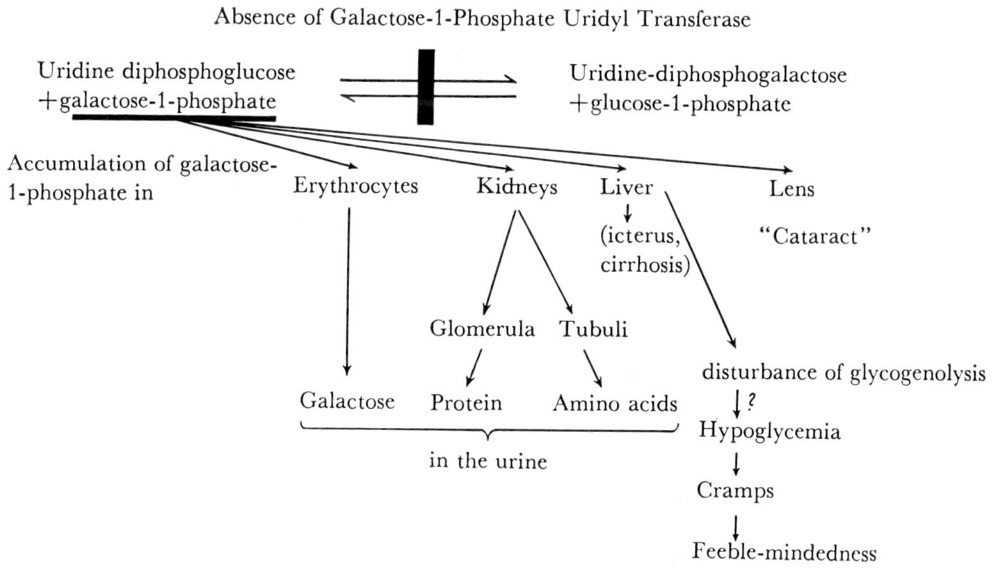

Absence of Galactose-1-Phosphate Uridyl Transferase

Uridine diphosphoglucose   ⟶   Uridine-diphosphogalactose
+galactose-1-phosphate   ⟵   +glucose-1-phosphate

Accumulation of galactose-
1-phosphate in   Erythrocytes   Kidneys   Liver   Lens
                                                              ↓
                                                       (icterus,   "Cataract"
                                                        cirrhosis)

                              Glomerula   Tubuli
                                  ↙   ↘   disturbance of glycogenolysis
                  Galactose   Protein   Amino acids   ↓ ?
                                                              Hypoglycemia
                        in the urine   ↓
                                                              Cramps
                                                                 ↓
                                                       Feeble-mindedness

FIG. 39.—Primary defect in galactosemia, and symptoms derived from it

symptomatic features in common. The characterization of a disease as "general mesenchymal inferiority" does not give much information. The designation "ectodermal syndrome" also tells nothing.

In the older literature, different syndromes whose individual signs apparently exist together without any internal connection frequently have been traced back to linked genes, that is, to several non-allelic genes which lie close together on a chromosome. For example, it has been stated that the Bardet-Biedl syndrome, in which polydactyly, obesity, feeble-mindedness, and retinitis pigmentosa occur jointly, is due to several individual genes. No plausible reasons have been cited, up to now, why such a syndrome would be caused by linked genes; instead, everything points to the explanation that, in the case of pathological syndromes with multiple symptoms, one is dealing with polyphenic effects of a single gene. In order to understand the reasons for this we must discuss linkage in more detail.

Results of location of several genes in one
chromosome. Linkage and interchange of factors
by crossing over

By means of linkage studies in Drosophila and, in recent times, also in the mouse, it has become possible to determine to a large extent the linear arrangement of individual gens on specific chromosomes. There are chromosome charts for Drosophila flies and mice on which the positions of the individual genes with respect to each other are indicated. For decades, such a chromosome chart has been a tempting goal for human geneticists also. Compared to the enormous amount of thought, time, and money which has been expended on these investigations, the result can only be described as meager. Genes are called linked when they lie on the same chromosome. Of course, acording to this definition, all X-chromosomal genes are linked to each other; thus the gene for agammaglobulinemia is linked to that for protanopia and to that for hemophilia. But this does not mean in any way that these three anomalies occur together more frequently than is to be expected in a random distribution. Linkage means only that genes which lie together in one chromosome in an individual also remain together in his children more frequently than do non-linked genes, namely, in more than 50 per cent of the cases. Thus, in the general population there is no correlation between linked genes. Someone who is a bleeder is for that reason neither more nor less likely to be red-green color-blind than any random man. But someone who is both a bleeder and red-green color-blind has both genes on his single X chromosome. It is therefore to be expected that an equally hemophilic brother of this patient will also be red-green color-blind, since he has the same X chromosome as his brother. Apart from the sex-linked genes, there are known individual anomalies which are linked to blood group loci and some blood group loci which are linked among themselves.

The scheme of linked hereditary traits with which we became acquainted on page 16 made it apparent that each of two "linked" genes can also lie on a different chromosome of the homologous pair. They are then in a state of repulsion. If, for example, a hemophilic man with normal color vision has a brother who is red-green color-blind but healthy, each of them has inherited a different one of the two maternal X chromosomes. Thus, in the mother, one of the X chromosomes has the gene for hemophilia; and the other X chromosome, the gene for color blindness; they are in the state of repulsion. In the sons, both genes cannot occur together, if the possibility of an interchange of factors is disregarded for the moment.

Autosomal gene linkage is considerably more difficult to determine. Mathematical procedures for this purpose were developed chiefly by Smith and by Morton (1956). They cannot be explained here. Definite examples of autosomal gene linkage are those between the nail-patella syndrome and the ABO series of alleles, and between a form of elliptocytosis and the Rh series of alleles. In the nail-patella syndrome—which is sometimes called the Turner-Kieser syndrome but which should not be confused with the Turner syndrome of gonadal dysgenesis, stunted growth, and malformations—the characteristic features are dystrophic nails primarily of the three radial fingers, a tendency toward luxations of the kneecaps and of the head of the radius, stunted growth, and peculiar bone protrusions in the pelvic bones. The two pedigrees of Figures 40 and 41 show the linkage of the syndrome with the gene for blood group B in one case, and, in the other, with the gene for blood group O. The linkage with the dominant gene for blood group B is easily recognized; linkage with the gene for O can only be inferred indirectly, since this gene is often masked by its alleles A and B. The third family, finally, shows multiple crossing over. In this family, the syndrome originally accompanied gen B, but in one afflicted family member the linkage was dissolved and, through crossing over, a new linkage with O was formed which now, until a further crossing over occurs, will remain in existence in the descendants also. Crossing over between the ABO locus and the locus of the nail-patella syndrome occurs in about 10 per cent of the cases.

The linkage is closer the closer the genes are located to each other, and looser, the farther apart they are. For two genes which lie at opposite ends of a chromosome, the probability is rather great that crossing over between the two will take place. This consideration is the basis for the calculation of the chromosome charts for Drosophila on the basis of the crossing-over rate. In man it is practically impossible to demonstrate linkage unless it is rather close. The differences in inheritance between loosely linked and non-linked genes are too small to be capable of proof on the basis of the limited material usually available for statistical analysis. Cytological investigations on spermatocytes have shown that the chiasmata which are the morphological basis of crossing over are very frequent. Thus, in the largest autosomes, four or five chiasmata often occur at the same time. The high frequency of the chiasmata also explains why, in spite of intensive search, so few linked genes have so far been found in man. Even linked genes are often almost freely interchangeable. This, of course, diminishes the significance of the linkage studies.

If very loose linkage is difficult to demonstrate in man, one might think that very close linkage should be much easier to demonstrate. Here also, however, a limitation exists. If two genes are linked so closely that crossing over between them is never observed, then the decision becomes impossible whether one is dealing with two genes or with a single gene with a twofold effect. The gene has been defined precisely as a non-crossing-over unit, that is, as the smallest unit of a chromosome which cannot be divided into sub-units through an interchange of factors. One can regard the factor complexes of the Rh system as single genes in this sense. Crossing over between single factors of these mosaic-like antigen complexes has never been observed. In micro-organisms and in Drosophila, as a result of the disproportionately greater number of generations and individuals observed under controllable breeding conditions, the sharpness of division between neigh-

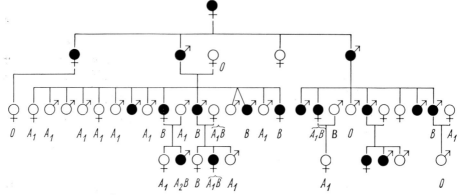

FIG. 40.—Excerpt from a pedigree with nail-patella syndrome. Only those sets of siblings in whom the syndrome occurred are shown in entirety. Healthy children of healthy persons, and healthy children of patients whose blood group was not determined, are omitted in part. Linkage without crossing over. (From Renwick and Lawler.)

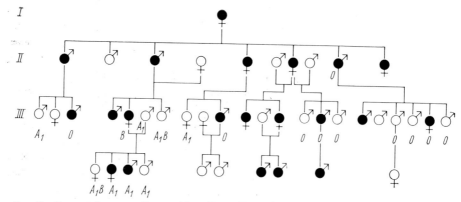

FIG. 41.—Excerpt from a pedigree with nail-patella syndrome. Linkage with blood group O. The female patient III₅ must have blood group genotype BO, since she has children of phenotype A₁. If the pathological gene were linked with B in her, her oldest daughter would have to be afflicted, but the next two children would be healthy. IV₂ and IV₃ must be of genotype A₁O. Since their father is healthy, but they have A₁ from the father, the family contains no recognizable case of crossing over. (From Renwick and Lawler.)

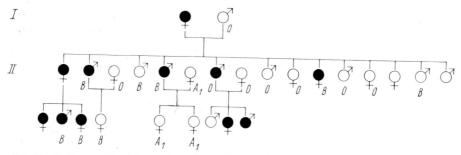

FIG. 42.—Pedigree with nail-patella syndrome. I₁ must have had blood group genotype BO, since she had children of phenotype O as well as phenotype B from a man with phenotype O. In generation II, the linkage between the syndrome and the gene for B is broken three times, namely, in II₄, II₇, and II₁₅. III₄ also has B without the syndrome, owing to crossing over. No conclusion can be drawn regarding linkage or crossing over in III₈ and III₉, since both parents had phenotype O, that is, they were homozygous OO. This pedigree alone would not make it possible to recognize linkage. (From Renwick and Lawler.)

boring non-crossing-over units is considerably greater than in man. In experimental genetics, therefore, even immediately adjacent genes which for many years had been regarded as genuine alleles often are finally shown to be "pseudo-alleles." In man, it is more practical to consider the Rh-factor complexes as a single series of alleles. This corresponds to the scientific principle that a theory should not contain more assumptions than are required for the explanation of the observations.

From a mnemotechnical point of view, to be sure, the interpretation of the Rh-complex antigens as products of closely linked genes has great advantages for everyone who does not work daily with this complicated system. Essentially, five different antibodies for demonstrating the Rh properties exist. At the suggestion of R. A. Fisher, who sees in these five antigen properties, which can be demonstrated with specific antibodies, the gene products of alleles on three different, closely linked loci, one calls the hypothetical genes c, C, e, E, and D. Here c and C, as well as e and E, are meant to be alleles. Fisher also postulates an allele d for D, although an antigenic property d has not been demonstrated up to now. The antigenic property D is the most important from a practical point of view because it is originally responsible for the largest number of cases of hemolytic illness of the newborn ("fetal erythroblastosis"). This hemolytic disease of the newborn is based on the fact that the mother can form antibodies against the antigens of the red blood corpuscles of the child, which the mother herself does not possess. The most widespread and strongest antigen of this type is D (Rh_0 in Wiener's nomenclature). Persons with the antigen D are usually called Rh-positive, and persons who do not have the antigen, Rh-negative. Thus, the chief danger is to the Rh-positive children of those Rh-negative mothers who are "isoimmunized" and therefore have anti-D in their serum. The transfer of maternal antibodies to the fetus leads to damage of the fetal erythrocytes and thus to hemolytic anemia. Until birth, this damage can largely be compensated by two special conditions of fetal life, so that severe anemia and hydrops result before birth only in rare extreme cases.

First, the fetus is protected because it can discharge the bilirubin which originates from the increased decay of its erythrocytes into the maternal circulatory system. The maternal liver easily handles the excretion of bilirubin. When the child has been born, however, and thus has been separated from the maternal circulatory system, the excretion of bilirubin has to be taken over by the child's liver. But the liver of the newborn does not yet possess the capacity, required for bilirubin excretion, of combining the bilirubin with glucuronic acid, so that a dangerous degree of jaundice sets in rapidly. The effects of the raised bilirubin level constitute the main danger to the newborn. This can lead to "kernicterus," bilirubin impregnation, and severe damage of cells of the nervous system.

The second protective mechanism which shields the fetus from anemia is the compensatory hyperplasia of the blood-forming tissue which leads to the liberation of numerous erythroblasts into the blood, that is, to erythroblastosis. This compensatory hyperplasia of the blood-forming tissue is dependent on the oxygen tension in the blood. The lack of oxygen in fetal blood is a powerful stimulus to blood formation. With birth and the commencement of respiration, the oxygen content in the infant's blood increases rapidly, but erythropoiesis is thereby suddenly suppressed and anemia sets in. Fortunately, it is possible today in most cases, by means of properly timed exchange transfusions, to protect the children from severe brain damage or death. Required for this is a determination of the Rh factor D in all pregnant women, and a determination of the maternal antibodies during the last months of pregnancy in all Rh-negative women. If this examination has not been performed, it is still possible immediately after birth to determine by means of the Coombs test in the blood of the umbilical cord whether the infant's erythrocytes possess antibodies. The frequency of erythroblastosis in children is approximately 0.3–0.5 per cent of all births. Thus, this is a problem of great practical significance. The present-day situation—in which, because of financial considerations, the required tests are not yet being carried out everywhere so that a large number of avoidable cases of death and avoidable physical and mental crippling persists—is urgently in need of reform.

If only anti-D serum were available, the genetic situation could be represented simply. We would then have a homozygote frequency for dd of 0.16, corresponding to a gene frequency for d of $\sqrt{0.16} = 0.40$. The frequency of D would amount to $1 - 0.40 = 0.60$. The homozygote frequency of DD would be $0.60^2 = 0.36$. The heterozygote frequency turns out to be $2 \times 0.6 \times 0.4 = 0.48$. Although 84 per cent of the total population is Rh-positive or D-positive, among Rh-positive persons the heterozygotes predominate over the homozygotes in the ratio of 4:3. Of course, this is also of practical significance, since it shows that one may expect that the Rh-positive father of a child will frequently be heterozygous. If the father is heterozygous, the chance that a second child is likewise afflicted is only 50 per cent.

The situation of the Rh-rh system is actually more complex than would be found merely through investigations with anti-D serum. The property of being D-positive or D-negative is always inherited en bloc together with different combinations of the properties c and C as well as e and E. Since c and C as well as e and E are alternative properties of which an individual always possesses one or the other or both, it has been assumed that each pair of antigens has as its basis a pair of alleles. One pictures these two pairs of alleles as closely linked with a third pair of alleles, DD, Dd, or dd. But with this, too, the complications are not yet at an end. A fourth pair of alleles, Ff, has been added and perhaps a fifth, and the individual loci may be occupied by additional alleles. Wiener, however, supports with emphasis the concept that only a single locus is involved, which may be occupied by numerous different alleles, each of which may determine several antigenic properties. At the level of the antigens, which evidently are very close to the primary gene products, Wiener's alleles would thus be polyphenic. The difference between the two concepts can be made clear by a comparison which Pontecorvo has employed for explaining the difference between the classical and the modern concept of chromosomes. The classical concept of chromosomes, to which Wiener's concept corresponds, would be comparable to a sentence composed of Chinese symbols standing next to each other as complex mutually independent elements. The modern concept of chromosomes, on the other hand, would be more comparable to handwritten lines of a modern letter script in which the demarcation of the individual letters and

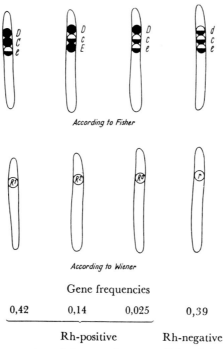

According to Fisher

According to Wiener

Gene frequencies

0,42   0,14   0,025   0,39

Rh-positive   Rh-negative

Fig. 43.—Schematic representation of the most frequent alleles of the Rh system, according to the concepts of Fisher (*above*) and Wiener (*below*). The great difference in frequency between R^0 and r is an indication against linked genes as the basis of individual antigenic properties.

of the individual words may be problematic but in which there are—besides the smallest units, the letters—larger functional units such as syllables and words. In the schematic representation of Figure 43, the two concepts of the genetic basis of the Rh system are compared. A decision in favor of one or the other is not yet possible at this time. In the end, this is really only a question of nomenclature. Everything depends on the way in which the gene is defined.

Speaking against linkage is the fact that the antigen properties c, C, D, e, and E show distinct correlations with each other within the population. The frequency of the individual antigens thus does not permit a calculation of the frequency of combinations, since these combinations do not follow a random distribution. For example, almost all D-negative persons are c-positive, e-positive, but C-negative and E-negative; thus, according to Fisher, they have the genotype cde/cde. Of the D-positive persons, on the other hand, only about 1/40 are at the same time c-positive, e-positive, C-negative, and E-negative. If the possibility of crossing over existed, it should have been expected that cc and ee would not occur more frequently in Rh-negative than in Rh-positive persons.

In the case of the genes for the red-green color defect, it has not yet been decided whether these belong to a single series of alleles or whether there is a locus for the protan series and another, also X-chromosomal locus and therefore linked to the first, for the deutan series. An observation favoring the two-locus hypothesis is that doubly heterozygous women who have one gene for red color blindness on one X chromosome and one gene for green color blindness on the other have normal color vision. The simplest expla-

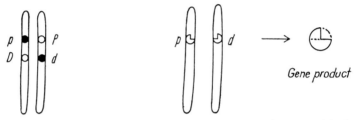

Fig. 44.—Doubly heterozygous state for protanopia and deuteranopia. *Left:* With the assumption of linked genes. Normal condition due to the two corresponding normal alleles. *Right:* With the assumption of multiple allelism (one-locus hypothesis). Mutual compensation of the genes in the gene product.

nation for this would be, of course, that these women possess the normal allele for each of the two pathological genes. It could also be true, however, that two alleles compensate their defect reciprocally according to the scheme of Figure 44. Curt Stern (1958) has pointed out, in favor of the one-locus hypothesis, that men who have red and green color blindness at the same time are much more rare than would be expected with a random distribution. Indeed, one would have to assume that every eightieth man with a red-green color defect has a protan as well as a deutan disorder. However, if both anomalies are brought about by mutually independent mutations of the same gene, then these two disorders could not combine in an individual with only one X chromosome. So far, there has been only a single instance in which a composite protan-deutan disturbance was observed in a family. This could be due to a special mutation of the one gene. This example will be discussed in more detail because it is intended to show the way in which linkage studies could bring about a solution of the problem, and in the following manner: The linkage of the two genes no longer needs to be proved. Both are X-chromosomal and thus linked by definition. But linkage studies could give us information about whether crossing over occurs. If crossing over is observed, this would prove that we are not dealing with alleles. A woman who has inherited from the father's side a gene for protanopia, and from the mother's side, a gene for deuteranopia, could, without crossing over, only have sons with either red color blindness or green color blindness. With crossing over, however, she could also have sons with normal color vision or a combination of red and green color blindness. To be sure, here again, very close linkage (or pseudo-allelism) could not be distinguished from multiple allelism. Close linkage of the genes for red and green color blindness would also be of interest theoretically, since it would favor the idea that in man, too, neighboring genes may have related functions.

Verschuer (1938) and Rath were the first to observe crossing over between X-chromosomal hemophilia and red-green color blindness. According to observations available today, this crossing over occurs in about 10 per cent of the cases. Between X-chromosomal muscular dystrophy and red-green color blindness, crossing over was found in five out of twenty cases. Studies on hemophilia B are not yet available. They could be of significance beyond their special interest for genetics. If hemophilia A and hemophilia B were indeed pseudo-alleles, as Vogel (1955) suspects, this would speak in favor of a functional relationship between anti-hemophilic globulin A and anti-hemophilic globulin B. However, if the two exhibit distinctly different crossing-over rates with the gene for red-green color blindness, it would have to be assumed that the two genes lie on the same chromosome only accidentally. Linkage studies have shown that dominantly inherited elliptocytosis is linked with the Rh factors in some families, but not in others. To the geneticist, this means that elliptocytosis is not a homogeneous disease. The clinician should now also attempt, in collaboration with biochemists and hematologists, to detect the difference between the two kinds of elliptocytosis by chemical and clinical means.

Heterogeneity of hereditary diseases and its significance for the method of classification (nosology). Different genes can cause similar diseases

Heterogeneity of hereditary diseases has been encountered on many occasions. This is the fact that a disease which is treated in medical textbooks under a single title often is not homogeneous when looked at genetically. Scientific advances lead to more and more division of the conventional collective groups into delimited individual types of diseases. The separation of hemophilia A and hemophilia B, or of the adrenogenital syndrome with high blood pressure from the type without high pressure, means a step forward. It is true that some authors find it disturbing that progressively more numerous subtypes and new syndromes are delimited; they attempt in the face of this to get along with large collective groups for which they coin broad covering terms. From the genetic point of view, such covering terms are a misleading step backward. Human genetics deals predominantly with well-defined anomalies which have no transitions from one into the other. Where presumable transitions are found, they often are either new types or chance combinations. Unfortunately, unusual combinations of rare diseases are described preferentially in the medical literature, without regard to the question whether these cases offer any new information. Also, the concept of the "relationship" between different syndromes is unclear. The hereditary syndromes, and the genes which form their basis, do not maintain relationships to each other. The actual basis of the so-called relationships usually consists in rather superficial symptomatic similarities or common pathogenic mechanisms. Rare syndromes are often described by authors who have seen the syndrome in question for the first time. The diagnosis made then is not always appropriate, and the description is not always adequate to make possible a correct diagnosis later. This is a contributing reason why the superficial case literature is teeming with transitional forms, while these hardly occur in thorough investigations.

Thus, the situation of the syndrome concept in the field of skeletal disorders may still be compared in some respects to the situation in the field of coagulation disorders as it presented itself twenty to thirty years ago, when one had to be satisfied with few clinical data and data from hereditary biology, with the anamnesis, and a few laboratory tests such as the determination of the coagulation time, the bleeding time, the retraction, and the number of thrombocytes. It is not surprising that the concept of a general hemorrhagic diathesis, which could manifest itself sometimes in one and sometimes in another way, was still defensible at that time. With the present more advanced methods of coagulation analysis, however, numerous individual types can be clearly distinguished from one another. If we knew the biochemistry of bone and cartilage as well as we know the mechanism of coagulation, we probably would have a better insight into the presently confused enchondral dysotoses and would not have to retreat to using a diffuse collective term. If genetic investigations are also called upon, as is done in exemplary fashion particularly in Maurice Lamy's school in Paris, and if, at the same time, the symptomatics of bone disorders are studied exactly, it is already possible to make a satisfactory tentative classification of enchondral dysotoses at the present time.

It cannot be expected that all the different syndromes be separately enumerated and discussed in a short textbook on a specialized medical field. But it would be desirable at least to be able to recognize from the titles that groups of diseases are being dealt with. Thus, a book on neurology should no longer contain the chapter heading "Progressive Muscular Dystrophy," but rather "The Progressive Muscular Dystrophies," and a textbook on pediatrics, not "Glycogen Storage Disease" but "The Glycogenoses." It makes little sense to collect clinically heterogeneous diseases of both a hereditary and a non-hereditary nature into one group, as is customary, for instance, under the name of phacomatoses. The term phacomatoses applies to a group of diseases which produce symptoms on the skin and in the nervous system, namely, neurofibromatosis, tuberous sclerosis, and Sturge-Weber's disease.

These considerations about nosology are valid only for hereditary diseases which are essentially determined by one gene, that is, whose etiology is known. In all exogenous diseases and in those determined by numerous factors of endogenous and exogenous nature, the causes vary in their intensity, their composition, and the timing of their effect. Here, true continuous transitions are frequently found.

CHAPTER FOUR

MUTATIONS

General discussion of mutations. Gene mutations

We understand the word "mutations" to mean abrupt changes of the genotype, which can affect a single gene (so-called point mutation), a chromosomal section (deletion of a piece, chromosome breakage, inversion of a chromosomal section), the number of chromosomes (monosomic: loss of a chromosome of the diploid set; trisomic: doubling of a chromosome of the diploid set), or the entire set of chromosomes ("ploidic mutations," triploid or tetraploid chromosome set instead of the normal diploid set). In man, point mutations are known primarily and, in addition, a few examples of genome mutations (one excess chromosome: Klinefelter's syndrome; trisomic X chromosome: mongolism; loss of a sex chromosome: Ullrich-Turner syndrome), as well as examples of suspected chromosomal mutations (translocation of a chromosome fragment). Because of mutations, the continuity of inheritance is suddenly interrupted. The participation of a mutant germ cell in the conception of an individual leads to genetic characteristics not present in the parents. The mutant genes or "mutants," however, are passed on to further descendants in the same manner as are other genes. At some time, every normal and pathological gene was first produced by mutation from another gene. The difference between pathological and normal here is not a difference in the nature of the gene, not even a fundamental difference in the primary gene effect. Instead, the difference between a normal and a pathological gene must be looked for simply in the fact that the effects of the normal gene are fitted by selection into a total genotype which is advantageous for the survival of the species, while those genes are called pathological whose effects disturb the joint action of the genes of the normal genotype which has been selected as a result of adaptation. Normal alleles provide the organism with a greater probability of preservation and therefore are passed on to more generations than are pathological genes, which diminish the probability of survival and of the propagation of their carriers and which therefore are eliminated again and again. The decisive reason why a new mutant usually leads to disease is that the genes which persist in the population are attuned to one another in their effect. Each alteration of a relatively harmonious equilibrium will in all probability upset this equilibrium.

The recognition of a new mutation is simple only if the mutant gene is recognizable with certainty in the heterozygous state. For the recognition of a dominant mutation the following conditions must obtain:

1. The gene must have regular dominance, that is, it must always lead to a specific characteristic. If the gene can be present without being recognized, one never knows whether a suspected mutation may not be due only to the manifestation of a gene which was already present among the antecedents.
2. There must not be any recessive genes which have an effect like that of the dominant gene.
3. There must not be any exogenous factors which can copy the effect of the dominant gene (so-called phenocopies, see below, pp. 148 ff.).
4. The parents of individuals with the characteristic produced by the new mutation must be available for examination, or at least reliable information about them must be at hand.
5. The parents reported must really be the biological parents. Non-marital or extra-marital conception is a possibility which the human geneticist must always take into account. It also happens that an adoption about which the parents have not informed the child may not be communicated to the physician either. Therefore, making certain of the biological father by means of blood group investigations, etc., is desirable in order to confirm a mutation.

One can assume regular dominance of a gene if 50 per cent of the children of carriers of a trait with healthy marriage partners again have the trait. This must also be true for the children of patients who are diseased for the first time in a family. If the remaining conditions are also fulfilled, one can look upon the cases of a disease occurring for the first time in the family as new mutations. Usually, the mutation rate does not refer to the population number, but to the number of genes in an allele series in the population. Since every individual has two genes per locus, the mutation rate per gene corresponds to half the frequency of the newly occurring cases in the population. In the mature germ cells ("gametes"), each gene is present only once. Therefore the mutation rate also applies to the number of mutations per gamete.

Since the above-named conditions for the recognition of a mutation do not always obtain in an individual case, it is difficult to make a reliable direct determination of the mutation rate. Therefore, it is important to have available for the determination of the mutation rate an indirect method which is not dependent on investigation of the parents of the probands or on the reliability of their biological parentage. The indirect method starts with the consideration that the percentage of cases produced by new mutations among all cases of a dominant hereditary disease must be proportional to the degree of limitation of propagation by this hereditary disease. The more severe a hereditary disease is, that is, the more it diminishes the probability of propagation, the fewer cases can have been inherited from the parents' generation. In the theoretical limiting case of a dominant hereditary disease which would completely exclude propagation, all cases would necessarily have been produced by new mutations. However, in this case, the dominant nature of the ailment would not be susceptible to proof. Presumably, acrocephalosyndactyly is an instance of such a dominant hereditary disease. In this anomaly, one usually finds a high degree of syndactylies of all fingers and toes, so that even the nails may be fused into a single elongated nail plate ("spoon hand"), as well as a pronounced towering skull. The patients are so disfigured and disabled that they do not usually propagate.[1] The analogy of form with several known dominantly inherited anomalies of the skull and of the ends of the extremities is conspicuous. In contrast, no similarity exists to any known exogenous disturbances of embryonal development. Also, in acrocephalosyndactyly all reliable indications are missing that exogenous damage during pregnancy could be responsible. The fact is to be noted that acrocephalosyndactyly depends on the age of the father in the same way as do the new mutations in chondrodystrophy. Presumably both disorders are due to a special type of mutation.

On the other hand, mild dominant anomalies such as Pelger's nuclear anomaly of the leukocytes or elliptocytosis may be transmitted through numerous generations. In this case it is to be expected that most of the afflicted individuals will have inherited the pathological gene from an equally afflicted parent and that new mutations are responsible only for a very small percentage.

The indirect method of determination of the mutation rate makes the further assumption that the frequency of a hereditary disease remains constant. The frequency of a dominant hereditary disease depends, first, on the mutation rate and, second, on the reproduction probability of the patients. If one knows the frequency of the hereditary disease and the reproduction probability of the patients, the mutation rate can be calculated without requiring reliable information about the parents of the patients. Many dominant hereditary diseases cause a certain decrease in reproduction. Thus, Borberg found that the effective fertility (reproduction probability) of patients with tuberous sclerosis is reduced to 51 per cent of the fertility of the normal population. This does not depend on the biological infertility as such, but on the fact that, because of epileptic seizures and feeble-mindedness, the patients are often placed in institutions, or at least do not found families. With an effective fertility of 50 per cent, half of all genes for tuberous sclerosis would be eliminated in one generation, and in the next generation again half of the still remaining 50 per cent, or a total of 75 per cent, and so on, so that, in four generations, 93.7 per cent of all genes for tuberous sclerosis would have disappeared from the population. If it were not for the fact that new cases keep occurring owing to new mutations, tuberous sclerosis would have disappeared long ago from the earth.

In Borberg's material, 57 per cent of the cases of tuberous sclerosis were sporadic. This value corresponds approximately to that expected on the basis of the decrease in reproduction probability. If half the traits are eliminated

---

1 Two clear-cut cases in mother and son have recently been observed (Blank, 1960).

in one generation and the disease still maintains the same frequency, half the cases must come about through new mutations. Thus the direct and the indirect method for the determination of the mutation rate have shown satisfactory agreement. In the case of Recklinghausen's neurofibromatosis, Crowe, Schull, and Neel found a reduction of the reproduction probability to 52.7 per cent. This reduction was due to several causes. Among 223 patients, 19 had lethal or severe complications involving the brain or the spinal cord before the end of the thirtieth year of life. Since some of the patients were still very young, one had to consider that several among them, too, would develop severe complications before the thirtieth year of life. Another factor influencing the diminished propagation is the shortened expectation of life. Neurofibromatosis occurs in both sexes with the same severity. In the investigations of Crowe and his co-workers it turned out, in spite of this sameness of the disease in both sexes, that the effective fertility of afflicted men was reduced to 41.3 per cent, that of afflicted women to 74.8 per cent of the fertility of their siblings. In part, this difference was due to the fact that afflicted men married more rarely, but in part also to differences in marital fertility between the married male and female patients. Presumably the total reduction of the effective fertility is determined by a complicated interplay of biological, psychological, and sociological factors.

In polycystic kidney disease of adults, more severe symptoms usually occur only toward the end and after the conclusion of the reproductive period. Therefore, the effective fertility is not reduced so substantially in this case: according to Dalgaard, to 77 per cent. Presumably for this reason also, sporadic cases are more rare than in tuberous sclerosis or neurofibromatosis. However, since the diagnosis can be established with certainty only by roentgenological examination or autopsy, and since the parents of the patients frequently are no longer alive at the time of diagnosis, it is impossible to obtain reliable data about the frequency of isolated cases. The mutation rate is calculated by the indirect method according to the following formula: $u = \frac{1}{2}(1 - f)x$, where u = mutation rate, in mutations of an allele series per total number of genes of the same series in the population (i.e., per gamete); f = reproduction probability, relative effective fertility, that is, average number of children of patients in relation to the average number of children of individuals in the normal population (patients who died early are included); and x = number of patients per total population.

For the example of polycystic kidney disease of adults, Dalgaard found: $f = 0.77$; $x = 0.001$; thus $u = 1.15 \times 10^{-4}$. Of every 10,000 genes of the allele series, then, 1.15 would carry a new mutation for polycystic kidney disease. The frequency of polycystic kidney disease used as a basis was 1 per 1,000. Accordingly, for 2,000 genes of the allele series there would be one mutant, and for 10,000, a total of 5 mutants. Among 5 mutants, the mutation would have occurred for the first time in 1.15. Approximately every fourth case of polycystic kidney disease would be based on a new mutation. Table 19 gives some estimates of the mutation rate of dominant genes.

Table 19 cannot be regarded as representative of all dominant mutations. In fact, the calculation of the mutation rate presupposes a certain frequency of the anomaly and, further, a distinctly reduced effective fertility. There is a sizable number of dominant anomalies which, first, appear to be more rare than those cited and which, second, diminish the reproduction probability to a lesser extent. For these anomalies, then, a lower mutation rate would have to be considered likely. Belonging in this category are, for example, special types of syndactyly and polydactyly, of symphalangism, the nail-patella syndrome, and dysostosis cleidocranialis. Thus, the table gives maximum rather than average figures for mutation rates of human genes.

TABLE 19

MUTATION RATES OF AUTOSOMAL
DOMINANT GENES*

Anomaly	Mutation Rate
Chondrodystrophy	1.3×10^{-5}
Pelger anomaly	2.7×10^{-5}
Aniridia	5×10^{-6}
Retinoblastoma	6.5×10^{-6}
Neurofibromatosis	1×10^{-4}
Cystic kidneys	1.2×10^{-4}
Marfan syndrome	5×10^{-6}
Myotonic dystrophy	8×10^{-6}
Polyposis intestini	1×10^{-5}

* From Vogel, 1958a. In the case of mutation rates from different authors for the same disease, only one estimate is cited. The estimate for the Pelger anomaly is especially uncertain, since no control by means of indirect calculation is possible, as the feature apparently does not diminish the effective fertility by a measurable amount. In retinoblastoma, non-hereditary cases occur, which were separated by Vogel with the aid of an ingenious method. Vogel regards these cases as phenocopies. However, this is perhaps a case of mutations of somatic cells.

The calculation of dominant mutations can be done directly and indirectly. In the calculation of recessive mutations the direct path is barred. A recessive gene ordinarily is not recognized in the heterozygous state. Therefore, one can never determine when the first mutation took place among the ancestors.

The mutation rate of recessive genes can be calculated indirectly on the basis of the assumption that the frequency of recessive hereditary diseases also depends, first, on the mutation rate and, second, on the reproduction probability, and that an equilibrium exists between gene-producing mutations and gene-eliminating selection. It is further assumed that the recessive hereditary gene in the heterozygous state has no effect on the reproduction probability. In order to understand the relationship between mutation rate and frequency of a recessive gene which is lethal in the homozygous state but completely neutral in the heterozygous state, we shall start with a simply constructed model. In a fictitious population which is free of pathological genes, we may imagine mutations of a certain gene with a constant frequency of

1:100,000 genes. Gradually, more and more such genes accumulate in this population, until they become so frequent that the chance of the encounter of two mutant genes is equal to twice the mutation rate. At this moment, two genes always disappear owing to the death of an individual who has become homozygous, and two new genes occur through mutations. Before this frequency has been reached, genes would also have disappeared through selection, but these would not be as numerous as those created by new mutations. The moment of equilibrium would be reached as soon as the gene frequency amounts to 1:224. (Double mutation rate $2:100,000 = 1:50,000 = 1/224 \times 1/224$.)

Since every individual possesses two genes of each type, a gene frequency per total number of alleles in the population of 1:224 corresponds to a frequency of carriers of the trait of 1:112. On the average, a mutant gene would be transmitted through 224 generations before it encountered a like pathological gene. The recessive genes which have become apparent in homozygous individuals form only a minute fraction of the total number of pathological recessive genes in the population. The chance that one of the two genes of a homozygous patient has accidentally been produced by a mutation is minimal in this case. It is arbitrarily assumed here that the population mixes at random in the choice of marriage partners. In actual fact, there is always a certain percentage of marriages between closer or more distant relatives, which considerably raises the probability that rare recessive genes will become homozygous. For a calculation of the mutation rate of recessive genes it would really be necessary to know the frequency of consanguineous marriages in the past 100 to 200 generations. Calculations which start with the conditions in existence today give values which are too low. In former times, consanguineous marriages were considerably more frequent, since people lived in smaller settlements which were relatively secluded. Because of this, recessive hereditary traits bcame homozygous more frequently and consequently became subject to selection. Today we live in a period in which there is no longer an equilibrium between recessive mutations and their elimination as a result of their becoming homozygous. The recessive genes will increase further until equilibrium has been re-established at the new level of frequency of consanguineous marriages. Even though the decreasing frequency of consanguineous marriages in Switzerland and in some Scandinavian regions in the past few generations has already led to a noticeable decrease of recessive hereditary diseases, in the long run the former frequency of recessive hereditary diseases will again be reached. We are profiting at the moment merely from the increased elimination of recessive traits through the higher frequency of consanguineous marriages among our ancestors. However, seen at long range, the number of recessive hereditary diseases is determined only by the mutation frequency and by the reproduction probability. The frequency of consanguineous marriages is the determining factor only for the latent period between mutation and manifest hereditary disease.

A further uncertainty in the determination of the mutation rate of recessive genes lies in the questionable assumption that the recessive traits in their heterozygous state have no influence on health and thus on the reproduction probability. If the recessive genes in the heterozygous state involve even a small disadvantage or perhaps an advantage, all calculations of their mutation rates are invalidated.

According to all our information about heterozygous manifestations of recessive genes, it is quite conceivable that these manifestations may have an influence on the vitality of their carriers under particular circumstances. Even though this influence may be only a small one, it may still carry some total weight because recessive genes spend by far the largest portion of their gene lifetime in heterozygous persons. We have fairly definite information about the significance of a recessive gene for the survival probability of heterozygous carriers only in the case of the gene for sickle cell anemia. As we have seen (see pp. 25 ff.), this gene is responsible for the formation of an abnormal hemoglobin, hemoglobin S. If attention is fixed upon sickle cell anemia, the gene is recessive with partial manifestation of the heterozygous state, but, if one considers the biochemical anomaly, it is codominant. Heterozygous carriers of the gene show no symptoms of disease, aside from the tendency toward spleen infarcts in high-altitude flights and a frequent microhematuria. Homozygous individuals frequently die during childhood. Their effective fertility amounts to a total of only around 20 per cent. It is surprising that, in some regions of East Africa, Greece, and India, a considerable percentage—up to 40 per cent—of the population is heterozygous for the gene for sickle cell anemia. Let us assume for the sake of simplicity of calculation that a stable equilibrium of gene frequencies exists in the population, and that patients with sickle cell anemia have as many children as the remaining population. Then, if the heterozygote frequency is 0.40, the gene frequencies for the normal and the pathological gene would amount to 0.72 and 0.28, while the frequency of homozygotes would be 0.077, or about 8 per cent.[2] It will be seen at once that the assumption of a stable equilibrium of genotypes cannot be correct. If 8 per cent of all individuals are homozygous, then for every 40 per cent of heterozygous individuals, who carry only one pathological gene, there would be $2 \times 7.7 = 15$ genes in homozygous individuals. Fifteen of a total of 55 genes, or 27.3 per cent of all genes in the population, can become directly subject to selection. The selection is not 100 per cent effective for all homozygous individuals, but is effective enough so that only 20 per cent in each generation reach the age of propagation. At first, it was difficult to understand why the sickle cell gene does not become

[2] From the heterozygote frequency $2pq = 0.40$, the gene frequency is calculated in the following manner:

$$p + q = 1; q = 1 - p;$$
$$2p - 2p^2 = 0.40; \text{ dividing by } - 2 \text{ gives}$$
$$p^2 - p = - 0.20. \text{ Completing the square,}$$
$$p^2 - p + 0.25 = 0.25 - 0.20; (p - 0.5)^2 = 0.05.$$
$$p = 0.5 \pm \sqrt{0.05}; p_1 = 0.72, p_2 = 0.28.$$

considerably more rare in a few generations and finally occur so seldom that the elimination rate and new production by mutations are held in balance. To explain this, it was necessary to assume considerably higher mutation rates than all those which were known in man and in other animals. Investigations in the Belgian Congo have shown, however, that mutations at this high rate do not occur. The resolution of this question succeeded because of the proof by Allison (1954) that the sickle cell gene in the heterozygous state carries with it a distinct preservation advantage in regions where malaria is endemic. The plasmodium falciparum, the agent causing malaria tropica, is evidently adapted in its metabolic requirements to normal hemoglobin.

The malaria plasmodia are parasites of the erythrocytes, living intracellularly and feeding on hemoglobin. If the organism presents them with an abnormal hemoglobin, the plasmodia do not flourish as well as with normal hemoglobin. Of course, it could also be that the sickle cell hemoglobin gives the red blood corpuscles diminished resistance to the malaria plasmodia, so that the afflicted erythrocytes perish more quickly along with the parasites. At any rate, under conditions of chronic infection with malaria tropica, the abnormal hemoglobin S confers an advantage which appears in improved resistance to the disease and thus, finally, in higher effective fertility. In regions of endemic malaria, heterozygous carriers of the sickle cell trait have about a 17 per cent higher chance of reaching adulthood than do individuals with normal hemoglobin A. Hemoglobin S can be regarded as a natural malaria remedy, created by mutation and bred by selection in some populations, which often has lethal effects in the homozygous state but which, in the heterozygous state, provides such a great benefit that the severe disadvantages of the homozygous state are thereby balanced as far as the population as a whole is concerned.

In addition to the mutation rate and the effective fertility of the homozygotes, the reproduction probability of the heterozygotes also participates in the equilibrium. Selection can never lead to a state in which the entire population has the sickle cell gene, since the percentage of those dying of sickle cell anemia would then become greater than the percentage of those saved by the increased resistance to malaria. With the heterozygotes at such a distinct advantage, the mutation rate no longer plays any practical role in regard to the gene frequency. Regardless of whether the mutation rate is very low or very high, the genes for sickle cell hemoglobin could at first spread more and more in the endemic malaria region; if the mutation rate is very low, it would merely take longer, but the equilibrium state would finally be determined solely by the relative effective fertility of heterozygotes and homozygotes. If a population with a high sickle cell gene frequency is now transplanted to a country where malaria does not exist, the advantage of the heterozygotes disappears, but the disadvantage of the homozygotes remains, and the result is a gradual decrease in the gene frequency. This presumably is the case with the Negro population in the United States. One might believe at first that the frequency of the sickle cell gene among the Africans is not a racial feature but a gene frequency which has been bred under special environmental conditions and which is characteristic for a specific population. In this sense, however, in the end all racial features are nothing more than gene frequencies characteristic for specific populations, to be traced back to selection factors which we do not yet know in detail. Presumably a similar protective mechanism against malaria is responsible for the wide distribution, which is hardly less surprising, of the gene for thalassemia in Italy, Spain, Greece, the Near East, and Indonesia. The direct proof for this has not yet been found. Still, it is very remarkable that there is no population on the earth in a country free of malaria in which more than extremely rare cases of thalassemia occur, while thalassemia is frequent particularly in some parts of Italy, for example in Sardinia or in the Po Delta, which formerly were infested with malaria.

Among the remaining recessive hereditary diseases, cystic fibrosis of the pancreas is prominent in its frequency. Here, also, an unusually high mutation rate would be required in order to replace the loss of genes occurring through the deaths of homozygous patients. Here again, therefore, one could think of the possibility that the heterozygous carriers of the trait may possess an advantage. So far, no direct evidence for this conjecture exists. On the contrary, it appears that certain anomalies, especially of lung function, set in even in heterozygotes. But perhaps an advantage still exists under special conditions. This would need to be only very small, so small that the prospect of ever actually demonstrating it would be unlikely. Let us assume that heterozygotes for the gene of cystic fibrosis of the pancreas have greater resistance to infections of the breathing passages or of the intestines. In a population in which intestinal infections and diseases of the air passages are frequent causes of death, even a small advantage could keep the gene frequency high. For a gene frequency of 0.03, the frequency of heterozygous gene carriers of cystic fibrosis of the pancreas amounts to 0.058, and the frequency of the illness, to 0.0009. Of all the genes in the population, only approximately 3 per cent would be found in homozygous individuals. Thus, 3 per cent of all genes would disappear in one generation. If the genes in the heterozygous state, in turn, provided a reproduction probability which is increased by 3 per cent, the loss of genes could be balanced. It is probably beyond any practical possibility to prove a survival advantage of only 3 per cent.

Under the assumption that the frequency of consanguineous marriages remains constant, and that the heterozygotes have a propagation probability of 1.0, the frequency of autosomal recessive mutations is calculated according to the formula $u = (1 - f)x$, the symbols in the formula having the same meaning as in the formula for dominant hereditary diseases. The formula differs from the corresponding formula for dominant hereditary diseases only by the absence of the factor $\frac{1}{2}$. This difference is due to the fact that one gene disappears with one carrier of a dominant hereditary disease; with the genetic death of the carrier of a recessive hereditary disease, on the other hand, two genes disappear. In Table 20 some estimates of mutation rates of recessive genes are compiled, with all due reservations. A direct determination is possible only for the sickle cell gene which can be detected in the heterozygous state; the indirect determination contains the indicated sources of error.

Up to now, we have studied mutations of autosomal genes. The situation is somewhat more complicated for mutations of X-chromosomal genes. As the simplest example, we shall discuss the mutation of a recessive X-chromosomal gene which leads either to death before the beginning of the reproductive age, or to infertility. Such genes are the basis of agammaglobulinemia, the Wiskott-Aldrich syndrome (eczema, thrombopenia, bloody diarrhea, otitis, and lack of resistance), a type of Pfaundler-Hurler disease, and a type of progressive muscular dystrophy. The responsible genes are located on the X chromosome. Of all X chromosomes in the population, 2/3 belong to the female sex and 1/3 to the male sex. If it is assumed, to begin with, that mutations occur with equal frequency in the male and in the female germ cells,

TABLE 20

SOME MUTATION RATES OF RECESSIVE GENES

Anomaly	Mutation Rate
Albinism......................	2.8×10^{-5}
Ichthyosis congenita..........	1.1×10^{-5}
Epidermolysis bullosa.........	5×10^{-5}
Total color blindness.........	2.8×10^{-5}
Infantile amaurotic idiocy.....	1.1×10^{-5}
Cystic fibrosis of pancreas.....	1.0×10^{-3} ??
Sickle cell anemia............	1.6×10^{-2} ??
	(direct $< 1.7 \times 10^{-3}$)

then the chance that the first individual with such an X-chromosomal mutation is male is 1:3; and that it is female, 2:3. A third of all mutations of this type would occur immediately in men and thus become subject to selection. Two-thirds, however, would be masked at first in female carriers. The X chromosome of a carrier has an equal chance of being transmitted to a son or to a daughter. Thus, in the next generation, half the concealed mutants which occurred originally in female carriers would come to light in sons; the other half would again be transmitted to female carriers. In each generation, half the mutations would become subject to selection and the other half preserved. The genetic survival of an X-chromosomal lethal gene has a half lifetime of one generation. Of 100 X-chromosomal lethal genes occurring in one generation, 33.3 would immediately appear in males and thus be eliminated; of the remaining 66.7 per cent which would be present to begin with in female carriers, in the following generation 33.3 per cent could be passed on to carriers, and in the third generation, 16.6 per cent of these; in the fourth generation, 8.3 per cent; in the fifth generation, 4.16 per cent; in the sixth, 2.08 per cent; etc. Thus, in five generations, 98 per cent of the X-chromosomal recessive lethal genes would again have disappeared. Here also, an equilib-

rium between mutation rate and selection is established. If it is assumed that
an equal number of new mutations occurs in each generation, and that the
selection opposing the hemizygous carriers of the trait is complete in each
generation, the scheme illustrated in Figure 45 is arrived at.

The ratio of genes due to new mutations to those inherited converges
toward 3:4, whereas the ratio of cases of illness created by new muta-
tions to those created by inherited genes approaches $1:1 + \frac{1}{2} + \frac{1}{4} + \frac{1}{8}$,
etc. $= 1:2$. For every three patients, there are four genes present in female
carriers of the gene in each generation. This ratio deviates clearly from the
1:2 ratio (more exactly, $p:2p - p^2$, where, for a small value of p, essentially
$p:2p = 1:2$ results) which is prevalent in the population among genes which
are neutral or nearly neutral with respect to selection, such as the traits for
red-green color blindness. It readily becomes clear that a considerable difference in distribution for the two sexes must exist between lethal and neutral
X-chromosomal genes, if it is remembered that the male carriers of the gene
pass on all X chromosomes to their daughters. If the male carriers of the gene
do not propagate, of course, there will be fewer female carriers of this gene
than of a gene which does not diminish propagation.

X-chromosomal recessive hereditary diseases actually do not always pre-

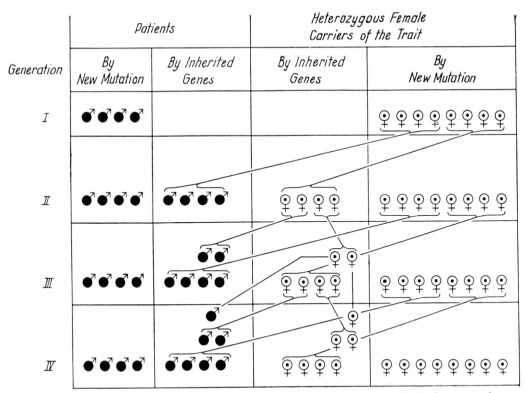

Fig. 45.—Diagram of equilibrium between mutations and selection for a recessive X-chromosomal
lethal factor.

vent hemizygous male carriers of the gene from propagation. Thus, some hemophiliacs propagate. The calculation of the mutation rate must then be modified correspondingly. If the hemizygous lethality of an X-chromosomal gene is 100 per cent, and if equilibrium between mutations and elimination exists, the number of new mutations would equal the frequency of carriers of the feature in the male population. On the other hand, for a sublethal gene which reduces the propagation probability to f, we obtain $u = 1/3 (1 - f)x$.

TABLE 21

DISTRIBUTION OF X-CHROMOSOMAL LETHAL GENES TO THE TWO SEXES

GENERA-TION	MUTANT GENES IN MALE ZYGOTES			MUTANT GENES IN FEMALE ZYGOTES		
	New Occurrence	Inherited from Female Carrier	Total	New Occurrence	Inherited from Female Carrier	Total
1........	1	....	1	2	....	2
2........	1	1	2	2	1	3
3........	1	1.5	2.5	2	1.5	3.5
4........	1	1.75	2.75	2	1.75	3.75
5........	1	1.875	2.875	2	1.875	3.875
∞.......	1	2	3	2	2	4

TABLE 22

ESTIMATES OF MUTATION RATES OF X-CHROMOSOMAL GENES

Anomaly	Mutation Rate	Place	Author
Hemophilia..................	3.2×10^{-5}	Denmark	Haldane
	2.2×10^{-5}	Switzerland	Vogel (1958a)
Dystrophia musculorum progressiva..................	9.5×10^{-5}	Utah	Stephens and Tyler
	7.4×10^{-5}	Northern Ireland	Stevenson
	4.3×10^{-5}	England	Walton
	4.8×10^{-5}	Germany	Becker and Lenz

If $f = 0$, that is, if the gene is 100 per cent lethal, the mutation rate equals a third of the frequency of patients per male population. In the scheme of Table 21, there are three new mutant genes for every three patients. The number of patients in the formula, however, is expressed relative to the male population, while the mutation rate is given in terms of the total number of X chromosomes in the population. Since women have two X chromosomes, but men only one, the total number of X chromosomes in the population is three times as great as the number of males in the population. In Table 22 some estimates of mutation rates of X-chromosomal genes are compiled.

The agreement between the different estimates is so good that one can assume the mutation rate of these genes to be a biological constant which is independent of the difference among populations. The mutation rate of "hemophilia" is actually composed of the mutation rates of hemophilia A and hemophilia B; these involve two genes which are independent of each other and probably not allelic.

Up to now we have assumed that mutations occur with equal frequency in male and in female germ cells. There is, however, some basis for the assumption that the mutation rate is actually higher in the male germ cells. The consequence of such a difference can be made clear most easily if one imagines what would happen if mutations were to occur only in the sperm cells. All mutant genes in the X chromosome would first reach female individuals, since the X chromosome of a sperm cell can join only with an X chromosome of the egg cell and thus always a female zygote is produced. Cases of illness which are based directly on a new mutation in the paternal germ cell would not exist. Only a mutation in the germ cells of the mother could lead directly to a diseased son. The mutations in the sperm cells would, at first, only lead to the creation of a female carrier, and to the occurrence of the illness only in her sons. If, indeed, the frequency of mutations in the male germ cells were considerably higher than in the female ones, the percentage of cases, among all cases of disease, which can be traced back to new mutations would have to be considerably less than with an equal mutation frequency in both sexes. A deviation of the observed figures from the conditions to be expected for an equal mutation rate in both sexes has actually been found by several authors for hemophilia and for X-chromosomal muscular dystrophy. However, the interpretation is not certain. The deviation could also be due to the more frequent detection of familial cases than of sporadic ones.

Apparently not all X-chromosomal genes have the same mutation rate. X-chromosomal agammaglobulinemia is definitely a rare hereditary disease, and so is the X-chromosomal Wiskott-Aldrich syndrome of thrombopenia with eczema and decreased resistance, and the X-chromosomal Pfaundler-Hurler disease. Reliable statistical investigations are not available, but there seems little doubt that these anomalies are definitely less frequent than X-chromosomal muscular dystrophy and hemophilia. Probably a difference in mutation rates exists also between the two types of hemophilia, A and B. Although hemophilia B tends to be less severe than hemophilia A, it is distinctly less frequent. This must be due to the lesser mutability of the gene for hemophilia B.

At any rate, it should be made clear that X-chromosomal recessive hereditary diseases often also occur sporadically, partly as a result of new mutations, partly because the gene inherited from the mother may by chance have reached only a single son, a fact to be expected frequently in families with a small number of children. Among three cases of an X-chromosomal lethal ailment, one has been produced by a new mutation in a maternal germ cell, while the other two have been transmitted by the mother as carrier of the gene. But of the two latter cases, on the average, one comes from a new mutation in a germ cell of the grandparents, and the second from the grandmother as carrier. At least a third of all cases must therefore be sporadic, another third can occur familially only in a single generation, and only in the remaining third is familial occurrence in several generations possible. Even if familial occurrence is possible, it is not always realized if the number of children in the families is small. The share of sporadic cases would have to be still greater for an X-chromosomal hereditary disease which is always lethal in the male sex and which can at least greatly diminish the viability in the female sex, as was discussed for incontinentia pigmenti Bloch-Sulzberger. In this anomaly, health can be affected to such a high degree by spastic paralyses, severe visual disturbances, or epileptic seizures, that marriage cannot even be considered. The frequency of sporadic cases, therefore, is not an argument here against a purely genetic cause.

Following the general discussion of inheritance of X-chromosomal recessive lethal genes and the relative contribution of new mutants among them, we can now approach the question whether X-chromosomal lethal factors can be made responsible for the higher embryonal and fetal mortality of the male sex. This question may be answered relatively simply by studying the sex ratio of the siblings of products which have died in the embryonal or fetal stage, as well as the sex ratio of the siblings of their mothers. If a woman has an X-chromosomal gene which leads to intra-uterine death in the hemizygous state, she will produce four types of zygotes with respect to this lethal gene: male lethal zygotes, healthy sons, normal daughters who are free of the trait, and, finally, daughters who again are carriers of the lethal gene. Of the children who can develop to the full term of the pregnancy, two-thirds are girls and only one-third boys. If the woman has received the lethal factor through a mutation in a germ cell of her parents, she is a sporadic carrier, and the sex ratio of her siblings will be normal. If the mother of the woman was already a carrier, however, the 2:1 ratio of sisters to brothers will also prevail among the siblings of the woman. Among the children and among the siblings of women with repeated male abortions, one would expect a predominance of the female sex. Along with the high frequency of spontaneous abortions, and with the predominance of boys among them, one would expect further that a statistical analysis of the sex ratio in random families would show an accumulation of families having only girls. However, such an accumulation does not exist. If the ratio of girls to boys in sets of siblings with recessive X-chromosomal lethal genes is 2:1, the chance that all children in a family of four children are daughters would be $2/3^4 = 16/81$; that all are sons, $1/3^4 = 1/81$, or only a sixteenth of the former. In fact, detailed statistics have not provided proof of any greater predominance of families with daughters. Therefore, it must probably be assumed that recessive X-chromosomal lethal factors can be held responsible at most for only a small part of the higher male mortality. It is unknown to what other cause this higher mortality can be ascribed.

Somatic mutations

It may be assumed that mutations of the same type as those occurring in germ cells also take place in somatic cells. In general, the consequences of such somatic mutations for the individual will be small, since only individual cells and their daughter cells are involved. However, if a somatic mutation occurs early in the embryonal development, the mutant cell may become the mother cell for a large number of abnormal body cells. Somatic mutations have been discussed primarily in connection with certain localized forms of neurofibromatosis and with blood groups. Crowe and his co-workers found 4 cases among 223 patients with neurofibromatosis which were sporadic and which exhibited the skin symptoms and tumors of neurofibromatosis only in circumscribed regions. Presumably somatic mutations are also the basis of most sporadic cases of one-sided glioblastoma retinae. Studies of somatic mutations which concern the antigenic properties of the erythrocytes are still in their initial stages but promise to yield very precise quantitative information about the frequency of occurrence of somatic mutations. Atwood and Scheinberg found in 12 individuals that 0.5 to 11 per thousand of all red blood corpuscles had an antigen structure different from that of the remaining ones. Thus, in an individual of blood group A_1B and N, only the properties B and N remained in the abnormal cells. An individual who has become genetically non-homogeneous because of somatic mutation of some cells is called a mutation mosaic. It is conceivable that some malformations are due to such mutation mosaics. Somatic mutations involve only a part of the body cells. They can therefore be compatible with life, while a correspondingly severe germ cell mutation would lead to the death of the individual, since it involves all body cells.

There is a series of analogies between mutations and malignant tumors, which has led people to ascribe malignant tumors generally to somatic mutations. Like mutations, tumors usually occur "by chance," without recognizable cause, independent of the condition of the total organism and of the specific metabolism of the cell. Numerous chemical substances which can produce mutations also cause tumor formation. Ionizing radiations, also, can cause mutations as well as tumors. The malignant tumor property is inherited from one cell by all daughter cells, just as a mutant continues to be inherited. This cellular property is remarkably independent of widely different local and general influences. A tumor usually keeps its specific properties for years, in defiance of all therapeutic efforts, and transfers its specific properties by means of deposition of cells in organs having entirely different metabolic conditions. In the non-obligatory sense of analogy one is therefore probably justified in speaking of a "mutation" as the cause of a malignant tumor. Of course, this should not be taken to imply that the mechanism is identical with that of a gene mutation. In fact, it is rather unlikely. Rather the "mutation" which leads to the degeneration of the cell probably causes a disturbance of the propagation mechanism of the cell itself, and perhaps secondarily in the propagating genetic material. In a genetic mutation, in contrast, the propagation mechanism of the cell remains intact. Anomalies in the number of chromosomes and their structure, such as translocations, fragmentations, and ring chromosomes are not infrequently found in human tumors. It is not known how these gross changes of the genetic material are related to the primary cause, the tumor property of the cell.[3]

Chromosome and genome mutations

The mutations considered so far concerned individual genes exclusively. Mutations of this kind cannot be demonstrated morphologically in the chromosomes with the methods used up to now. In addition, there are mutations which involve not individual genes, but entire chromosome sections or chromosomes. The gross structural changes within a chromosome are called chromosome mutations. This may be a case of fragmentation in which the fragment which does not carry the centromere is lost in the next mitosis, since it no longer offers a place of attachment for the spindle fibers which distribute the chromosomes regularly between the two daughter cells. It is true that chromosome fragments frequently also attach themselves to other chromosomes and that they can then still be further preserved in the cycle of cell divisions (translocation). If, after a rotation through 180°, a chromosome fragment again heals at the place of breakage, we speak of an inversion. Except for some cases of presumed translocation and one case of chromosome breakage, these different chromosome mutations have not yet been observed in man, aside from malignant tumors, in which chromosome breaks and translocations are found rather frequently.

The next type of mutation is designated as genome mutation. Here, neither do genes undergo mutation nor has a disruption of continuity or a change in structure occurred within the individual chromosomes, but, instead, entire chromosomes have been lost or are supernumerary. If one chromosome of a pair is missing from the diploid set, we speak of a monosomic individual. If, on the other hand, a chromosome is present in triplicate rather than duplicate, we designate the abnormal individual as trisomic.

The largest unit in the organization of the genetic material is the genome, that is, the haploid chromosome assortment. If an organism has an abnormal multiple of its number of chromosomes, we speak of a ploidic mutation. Ploidic mutations play an important role in many plants. In contrast to genome mutations, in which the normal equilibrium of hereditary factors is usually disturbed quite considerably, so that viability and fertility are adversely affected, ploidic mutations in plants usually adapt well. So far, a ploidic mutation has been found in man only in aborted fetuses and in a case of a physically and mentally retarded child with syndactylies, micrognathia, and lipomatosis, who had a triploid set of 69 chromosomes with the combination XXY.

Genome mutations are of considerable theoretical interest because they emphatically demonstrate a new aspect of the organization of the genetic

[3] One special kind of tumor, chronic myelogenous leukemia, has been found to be related in most cases to a specific chromosome mutation, namely, deletion of part of the long arm of chromosome 21.

material, namely, that of genetic equilibrium. Organisms with genome mutations have no abnormal genes, and they are not lacking any normal genes; yet
they are usually morphologically abnormal, low in viability, and less fertile
than normal individuals. The cause of the diminished viability of trisomic
and monosomic individuals is probably a disorder of the normal equilibrium
of the genes. These relationships have been studied more closely, especially
in some plants (thorn apple, corn), and also in Drosophila and salamanders.

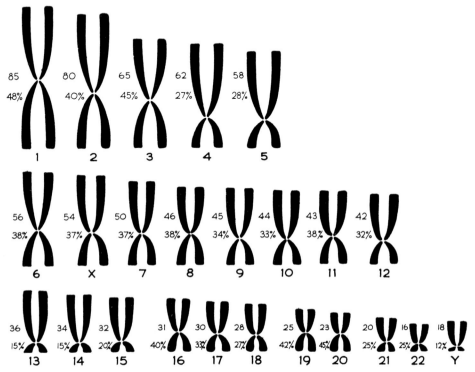

FIG. 46.—An idealized chromosome set, numbered according to the internationally adopted Den-
ver system. Only one of each pair is represented. The small figures beside each chromosome
indicate approximately the relative length of the whole chromosome and the proportion of the
total length occupied by the short arm. (From Lennox.)

The formation of monosomic and trisomic individuals can be traced
back to the occurrence of a disturbance of the normal chromosome distribution in the maturation divisions of the germ cells. In the male sex, the two
maturation divisions normally lead from the tetrad stage in rapid succession
to four haploid mature germ cells and, in the female sex, to one haploid egg
cell and three "polar bodies." Occasionally, the haploid formation may fail
to occur in individual pairs of chromosomes. If two homologous chromosomes, which normally should separate and reach different daughter cells,
remain together, we speak of non-disjunction. The result of such a non-
disjunction in one of the maturation divisions is that an egg cell has a super-

numerary chromosome in addition to the complete haploid chromosome set. If this abnormal egg cell is now fertilized by a normal sperm, a trisomic individual develops. One must probably assume, however, that most trisomic fertilized eggs are not capable of developing. At any rate, in man trisomy has been observed up to now only for three, possibly four, of the smaller autosomes and for the X and Y chromosomes. Trisomy for one of the smallest autosomes is the cause of mongolism. Mongoloid patients are always feeble-minded, though only rarely of the most severe degree of feeble-mindedness,

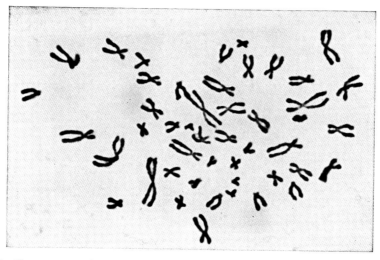

FIG. 47.—Chromosomes of an eleven-year-old patient with mongolism. (From Böök, Fraccaro, and Lindsten.)

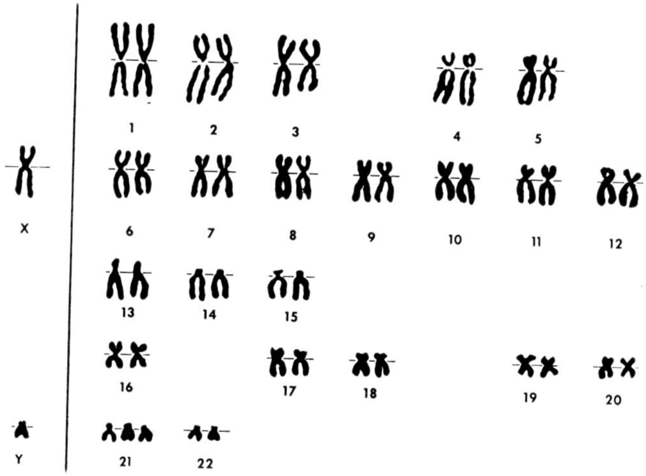

FIG. 48.—Karyotype of male mongoloid patient. (Courtesy of Dr. Pfeiffer, Münster.)

idiocy. Everyone who has some experience with mongoloids can easily recognize them because of the peculiarities of their facial expression, their posture, their speech, and a series of characteristic minor features, such as short or bent little fingers, four-finger fold (a crease which passes across the palm of the hand), certain patterns of the dermal ridges of the hand, and whitish spots in the iris. Heart defects are very frequent. The term mongolism was coined on the basis of an erroneous concept of a "reversion" to the Mongolian race. The designation is hardly appropriate in regard to the symptomatology, false in its basic hypothesis, and, furthermore, tactless. It should soon be replaced by a better term, such as "21 trisomy." For the time being, however, mongolism is still the only term that is generally understood. Therefore it will be kept here for the present.

Two facts are significant for the etiology of mongolism if, to begin with, one disregards the chromosomal findings:

1. The possibility of bearing a mongoloid child becomes increasingly frequent with increasing age of the mother.
2. A definite monozygotic pair of twins has never yet been observed in which only one partner was mongoloid while the other was healthy.

If one partner of a dizygotic pair of twins is mongoloid, the other is almost always healthy. There is only a very small familial accumulation of mongolism. Mongoloid women very rarely have children, probably owing less to a disorder in their potential fertility than to the high degree of feeble-mindedness. However, if they have children, these too may be mongoloid. So far, a total of five mongoloid and five non-mongoloid children of mongoloid mothers has been reported.

The dependence of mongolism on the age of the mother indicates that the disturbance of the distribution mechanism of the chromosomes (non-disjunction) occurs more frequently in an aging ovary than in a youthful one. The total frequency of mongolism lies between two and three cases per 1,000 births. For mothers below 20 years of age, the frequency is approximately 1 per 1,000, for mothers between 40 and 45 years, 15 per 1,000, and for mothers 50 years old, in the order of magnitude of 100 per 1,000 (i.e., 10 per cent). It has so far not been possible to demonstrate external causes leading to an increased incidence of mongolism. Also, the significance of genetic factors for the disposition to non-disjunction is evidently small.[4]

Trisomy of the X chromosome has been observed so far only in seventeen cases. The only constant characteristics of these women seem to be mental deficiency and the possession of double sex chromatin bodies. Perhaps such XXX women are not too rare.

One arrives at this conjecture by way of the following considerations: Abnormal men with the sex chromosomes XXY, which lead to the so-called

[4] Recently, a translocation of autosomes 21 and 15 has been demonstrated as a basis for familial occurrence of mongolism (Penrose, Ellis, Delhanty, 1960). Other cases are due to a 21/21 translocation, or, more likely, to reduplication of the long arm and loss of the short arm of a chromosome 21.

Klinefelter's syndrome, occur with a frequency comparable to that of mongolism—about 2 per 1,000 male individuals. It is assumed that the cause of Klinefelter's syndrome is to be found in a non-disjunction in the maternal egg cell, which leads to an abnormal ovum with a haploid set of twenty-two autosomes, but two X chromosomes. If such an abnormal ovum is fertilized by a sperm with a Y chromosome, a Klinefelter patient is created. The probability that the fertilizing sperm will contribute a third X chromosome, however, is not much smaller than that it will contain a Y chromosome. It is therefore to be expected that XXX women are approximately as frequent as abnormal XXY men. In Klinefelter's syndrome, we have assumed that the non-disjunction is also in the maternal ovum. The reason for this supposition is given by observations of Klinefelter patients with red-green color defi-

Fɪɢ. 49.—Karyotype of Klinefelter patient. (Courtesy of Dr. Pfeiffer, Universitäts-Kinderklinik, Münster.)

ciency who had inherited the X-chromosomal gene for their color sense disorders from the maternal side, while their fathers had normal color vision. If these XXY individuals with defective red-green vision had inherited a normal egg cell from the mother, but, from the father, an abnormal XY sperm created by non-disjunction, the father himself would have to have been color-defective. In one of three cases of Klinefelter's syndrome with defective red-green vision, the mother herself had defective red-green vision. There is no difficulty in understanding this. The mother evidently had a homozygous trait for the red-green color defect. Thus, her abnormal egg cells with two X chromosomes also must both have contained the gene for the red-green defect.

At first, it is more difficult to understand the two other cases of red-green color defect in Klinefelter's syndrome, in which the fathers had normal color vision, but the mothers were heterozygous for the gene. If we recall, however, that the two homologous chromosomes of a pair are always split length-

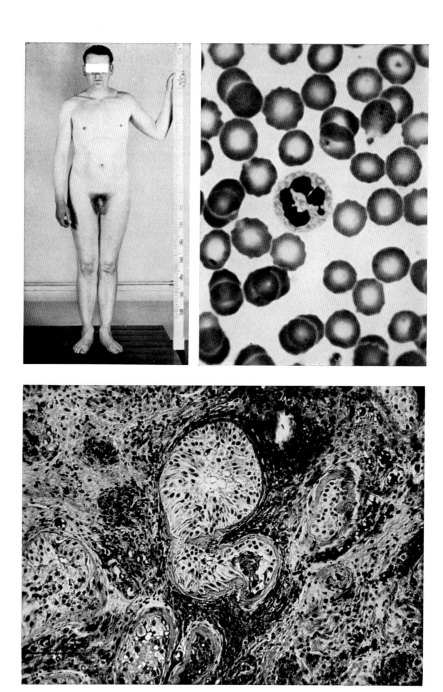

Fig. 50.—Patient with chromatin-positive Klinefelter's syndrome and protanopia

wise into two halves before the maturation divisions, and that segments can
be exchanged between the four strands of the tetrad stage through crossing
over, it is not hard to see how a heterozygous woman can produce homozygous non-disjunction ova in the first maturation division through crossing
over and non-disjunction. Of course, a non-disjunction would also be conceivable in the second maturation division, in which sister chromatids with
the homozygous pair of genes could again remain together, without the ne-
cessity of a previous crossing over. In large chromosomes, chiasmata, the
morphological basis of crossing over, are very frequent; therefore, for genes
which lie somewhat farther apart, one can in practice count on the ability of
the genes to be exchanged with complete freedom. Here, the linkage is
broken, not only in an exceptional case, but constantly. In Klinefelter's syn-

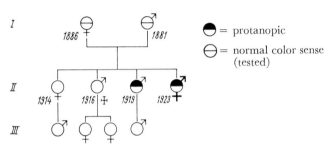

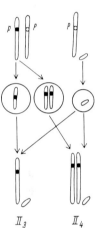

Fig. 51.—Family of the patient of Figure 50
with red-green color blindness and Klinefelter's
syndrome.

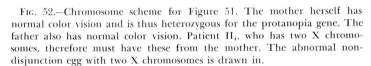

Fig. 52.—Chromosome scheme for Figure 51. The mother herself has
normal color vision and is thus heterozygous for the protanopia gene. The
father also has normal color vision. Patient II₄, who has two X chromosomes, therefore must have these from the mother. The abnormal non-
disjunction egg with two X chromosomes is drawn in.

drome, also, there is dependence on the age of the mother, but the frequency
increases after the mother's fortieth year of life only about three- to fourfold.
The curve of the maternal age distribution is so different for mongolism
and for Klinefelter's syndrome that the causes for the two types of non-
disjunction cannot be the same. There are some cases of patients on record
in whom a typical Klinefelter's syndrome and mongolism coexisted and who
had forty-eight chromosomes, including an excess sex chromosome as well as
an excess autosome.

Familial accumulation of Klinefelter's syndrome has not been observed as
yet, but twice it has occurred in each of a pair of monozygotic twins. Klinefelter patients seem to be always sterile, so that the chromosomal anomaly
cannot be passed on by them. In this sense, Klinefelter's syndrome is not
hereditary. If, however, one calls differences caused by the cell nucleus genetically determined, Klinefelter's syndrome is a genetically determined
disorder. Its basis is a disorder of the cell constituents which are normally

responsible for the transmission of hereditary traits, but which, only in this special case, are not passed on because of the disorder. This situation corresponds entirely to another genetically caused syndrome, each individual case of which apparently is based on a new genome mutation, this time on the loss of a sex chromosome.

In the Ullrich-Turner syndrome with gonadal dysgenesis, different malformations are found in combinations which are typical in a general way, but quite variable in individual cases, including especially webbing on both sides of the neck, funnel chest, stunted growth, cubitus valgus, coarctation of

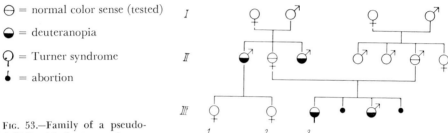

⊖ = normal color sense (tested)

◐ = deuteranopia

♀ = Turner syndrome

● = abortion

Fig. 53.—Family of a pseudo-girl (XO) with deuteranopia.

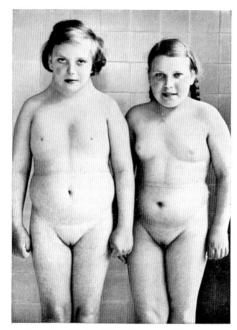

Fig. 54.—*Right,* III₃ of Figure 53. Next to her a second XO patient.

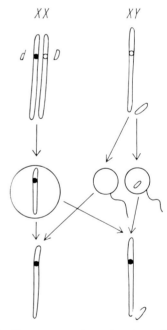

Fig. 55.—Chromosome scheme for Figure 53. Since the XO patient has received the X chromosome marked with the deuteranopia gene from her mother, she must have received the abnormal germ cell without sex chromosome from the father.

the aorta, fusions of vertebrae, multiple cell nevi of the skin, nail anomalies, receding chin, and high-arched palate. The external genitals are female. The gonads are present only in rudimentary form (gonadal dysgenesis). To be sure, there are also cases of gonadal dysgenesis which have nothing to do with the Ullrich-Turner syndrome. In the Ullrich-Turner syndrome, forty-five chromosomes are present, namely, the normal diploid autosome set, but only a single X chromosome. Since these patients are infertile, the genome mutation is eliminated in each case with its first carrier. If the fertilized ovum with an anomaly in the number of chromosomes divides, so that monozygotic twins are produced, both, of course, have the anomaly. Mongoloid monozygotic twins have been observed frequently; two pairs of monozygotic twins with completely identical anomalies have been observed for Klinefelter's syndrome, and one for the Ullrich-Turner syndrome.

The origin of the Ullrich-Turner syndrome, in which a chromosome is lacking, can also be explained by a non-disjunction. In this case, however, the cell which has received one less chromosome would have to undergo fertilization. While the only criteria available for mongolism point to an anomaly of the maturation divisions of the ovum, there are indications in the case of the Ullrich-Turner syndrome that the defect may be found in spermatogenesis. It has not been possible to prove any dependence on the age of the father or of the mother. Nevertheless, family examinations for red-green color-defective female patients with the Ullrich-Turner syndrome showed that the fathers of the patients had normal color vision, while the mothers were heterozygous for the red-green color defect. It follows from this that the X chromosome which is marked with the gene for the red-green defect must come from the mother. Thus the maternal ovum would have provided an X chromosome in the regular manner. The cell in which a sex chromosome was lacking must therefore have been the paternal sperm cell.[5]

Mutations and age of parents

The genome mutation which is the basis of mongolism is dependent in a very significant way on the age of the mother; the corresponding mutation which is the basis of Klinefelter's syndrome shows a lesser dependence on the maternal age; and the chromosome loss which leads to the Ullrich-Turner syndrome seems to occur independently of the age of the parents. At present, little definite information about the age dependence of gene mutations is available. Most of the dominant hereditary diseases on which data are available do not show any clear-cut influence. However, one exception in sharp contrast to this is chondrodystrophy. Newly occurring mutations are about ten times as frequent for fathers over forty years old as for very young fathers

[5] There is one report of monozygotic twins, as proved by reciprocal skin grafts and identical blood groups, of which one was a normal male and the other an XO-Turner individual (Turpin, Lafourcade, Chigot, and Salmon, 1961). This case, as well as the not too rare cases of XO/XX-mosaicism, points to mitotic non-disjunction after fertilization as a cause of chromosomal abnormalities.

(see Figs. 56 and 57). The mother's age is without significance here. The effect of paternal age is similar in acrocephalosyndactyly. Only it is not certain in the case of acrocephalosyndactyly whether it is a "dominant" hereditary disease, since the afflicted patients have no children (see p. 106). The mutation which leads to dominant chondrodystrophy must differ in some still unknown way from the other dominant mutations. This is an interesting finding, which shows that there are different kinds of point mutations. Chromosome studies on cell cultures have shown no morphological anom-

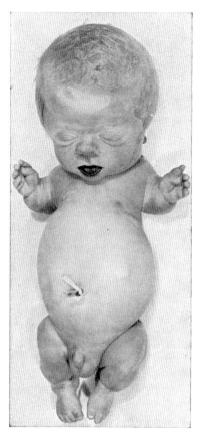

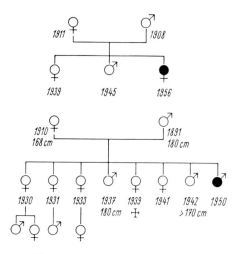

Fig. 56.—Chondrodystrophic stillbirth. Parents healthy. Father 42 years, mother 38 years old. Two healthy siblings.

Fig. 57.—Two typical families with sporadic chondrodystrophy in children of old fathers. (Own observation.)

alies in typical chondrodystrophy and acrocephalosyndactyly, so that a gross chromosomal aberration can probably be excluded from consideration. The number of different dominant anomalies is very large. It is possible that other dominant mutations, also, behave like chondrodystrophy. However, since reliable data are available only for a few, this question has not yet been resolved.

Production of mutations by ionizing radiations

It has been possible to demonstrate in all living beings whose mutation rate can be determined quantitatively that it is affected by ionizing radiations.

Since man does not occupy an exceptional position among living beings in regard to the structure of his hereditary material, one may expect in man, also, an increase of mutations due to roentgen rays and radioactive substances. It has not yet been possible to give a definite answer to the question of the order of magnitude of this increase. However, investigations have been carried out which at least provide criteria for establishing the radiation dose at which there is as yet no demonstrable hereditary damage to the direct descendants of those who have been irradiated.

First of all, studies on children of radiologists should be mentioned (Loeffler, Nürnberger, Macht and Lawrence, Crow). Even though no protective measures were taken during the initial period of radiological practice, and skin burns, radiation-induced carcinomas, and azoospermia were frequent occurrences among radiologists, no accumulation of hereditary damage could be discovered among approximately 7,000 children of radiologists. It can be assumed that the pioneers in radiology received a yearly gonadal dose of about 100 r while, under modern protection conditions, this value is perhaps still 1 r per year (Braestrup). Kaplan has reported on 566 children who were born of mothers who had received an ovarian dose of 65 r for treatment of their sterility. Here, also, no accumulation of malformations or of anomalies known to be hereditary was found.

The most extensive report on the possible genetic consequences of ionizing radiations is that of Neel and Schull concerning the children of men and women who were exposed in Hiroshima and Nagasaki to the rays of the atomic bomb explosions. A total of 36,219 of these children were examined at birth and at the age of nine months. The largest portion of the children came from parents who were at a distance of two to three kilometers from the center of the bomb effect at the time the bomb was dropped. Twenty-four hundred and sixty-six children came from parents at least one of whom had shown symptoms of radiation damage, such as loss of hair, petechiae, or gingivitis. The dose these parents had received was estimated at 300 to 400 r, that is, it was probably lower than the cumulative dose which the radiologists of the early X-ray period had received over the years. Among the children of Nagasaki and Hiroshima it was not possible to demonstrate an accumulation of congenital malformations or hereditary anomalies. In the conclusion of their report, Neel and Schull state that their results "remove the remote possibility of a conspicuous sensitivity of human genes to irradiation (i.e., marked mutability)."

Even if the investigations at Hiroshima and Nagasaki did not yield a direct determination of genetic damage, they nevertheless produced valuable indirect information on the increase of X-chromosomal lethal mutations as a consequence of the irradiation. If X-chromosomal recessive lethal factors occur in the maternal ova as a result of irradiation of the mother, the sex ratio of the products which develop to the stage of birth must shift in favor of girls. X-chromosomal lethal factors in the paternal germ cells, on the other hand, cannot reveal themselves in the children if they are recessive, since they are masked by the normal allele in the second X chromosome, which comes from the mother. In contrast, X-chromosomal dominant hereditary factors in the spermatozoa would lead to the death of female fetuses, and thus to preferential survival of male fetuses. Thus, for genetic reasons, one would expect among the children of irradiated women a shift of the sex ratio in favor of girls, but among the children of irradiated men, in favor of boys. This expectation was confirmed in the report on Hiroshima and Nagasaki, but the deviations were irregular and not statistically certain. A new analysis of the old data and of new data collected since, however, demonstrated statistically significant deviations which are reproduced in outline in Table 23.

If one assumes that the X chromosome contains 250 genes which can become lethal factors by means of mutation, one obtains a mutation rate per gene as a result of radiation exposure of 2.4×10^{-5} per 100 r. This would indicate that the radiation sensitivity of genes in man is approximately equal in magnitude to that of the mouse.

TABLE 23

PERCENTAGE OF MALE BIRTHS AMONG ALL BIRTHS FROM
PARENTS WHO SURVIVED THE ATOMIC BOMB EX-
PLOSIONS OF HIROSHIMA AND NAGASAKI

	Per Cent Male Births	Number of Children
Mother and father not irradiated......	52.1	46,166
Mother not irradiated; father 200 r.....	53.0	836
Father not irradiated; mother 200 r....	51.3	2,526
Father exposed; mother 200 r..........	51.0	870
Mother exposed; father 200 r..........	52.7	1,144

The predominantly negative results of all investigations concerning the influence of ionizing radiations on mutations in man should not be misunderstood as proof of the harmlessness of ionizing radiations. Every exposure to ionizing rays signifies a risk for the hereditary material which must be balanced against their possible usefulness. In principle, a single roentgen examination can cause a mutation; but such an event is extremely unlikely. For the individual with a hereditary disease who becomes ill because of such a mutation, however, it is no consolation that he owes his illness to an improbable event. The physician values the individual, and not so much the frequencies of diseases in the population. He must always keep in mind that X rays are a double-edged sword which may bring severe suffering in future generations still. However, with the widespread fear of X rays which exists in the public today, presumably more total damage to health results from neglected examinations than is balanced by the absence of genetic damage resulting from the avoidance of radiation.

One often hears the opinion expressed that damage from ionizing radiation cannot be recognized in the first generation, but only in the subsequent ones. This is based on a misunderstanding. It is correct that only a fraction of the total genetic damage is recognizable in the first generation, and that the total damage is distributed over many generations. Dominant mutations even occur most frequently in the first generation, as do X-chromosomal recessive mutations in the germ cells of women. But the recessive mutants, also, which are created by a single large dose of radiation, can lead to disease as early as the first generation. Within the population, there are numerous recessive genes of the same type as those which are radiation-induced. The probability that one radiation-induced mutant will encounter another of the same kind is not smaller in the first than in later generations; in fact, it is even somewhat greater. This probability depends on the frequency of the gene in question within the population. If the frequency of a recessive lethal gene in the population amounts to 0.005 (1:200), it will be transmitted on the average through two hundred generations, until it is eliminated by becoming homozygous. If the mutation rate of this gene is doubled by continuous irradiation of the population through many generations, the frequency of the gene gradually increases to 0.01. Then a new equilibrium between gene frequency and new mutation rate is established. The frequency of homozygotes originally amounts to $0.005 \times 0.005 = 0.000025$. The original mutation rate corresponds to the number of genes disappearing through the death of homozygotes, namely, 0.000025 or 2.5×10^{-5}. If the mutation rate is doubled owing to irradiation, an additional 2.5×10^{-5} enters in. In each generation, $2 \times 2.5 \times 10^{-5}$ of the individuals would possess the new radiation-induced mutation in the heterozygous state in one or the other allele. The original frequency of heterozygotes, however, amounted to approximately $2p = 0.01$, or 100 per 10,000. By means of an irradiation which doubles the mutation rate, a recessive gene with the frequency typical for recessive genes would increase in frequency only by about 0.5 per cent in each generation. Thus, more than two hundred generations would be required until the new gene frequency of 0.01 would be attained; and this does not take exactly two hundred generations, as would be thought at first glance, because the frequency with which the gene becomes homozygous, and thereby its elimination rate, increases only with rising gene frequency. This example is valid under the presupposition that all coming generations will also be subjected to the same radiation exposure that doubles the mutation rate. If, on the other hand, only one generation experiences such a radiation effect, the recessive mutations increase only by a minute fraction which it is impossible to determine, since their manifestation would be drawn out over numerous generations. We do not know what quantity of radiation leads to a doubling of the mutation rate. The estimates vary between 3 r and 150 r. For Drosophila, the value is approximately 30–50 r.

It has been shown for Drosophila that the mutation rate depends only on the total dose of ionizing radiations and not on the distribution of this dose over a longer or shorter period of time. However, investigations of radiation-induced mutations in the mouse have resulted in a more complicated picture. Under an acute exposure of short duration, the same amount of radiation leads to a larger number of mutations than if it is administered in fractionated doses over a longer period. This disproved the principle which had been accepted for a long time that induced mutations are dependent only on dose. The frequency of induced mutations depends also on the sex and the stage of development of the germ cells. A considerably larger number of mutations is induced in the spermatogonia of adult mice than in fetal spermatogonia or in oöcytes. T. C. Carter concludes from this that X-ray examinations of boys and young men cause the major portion of genetic damage. But this damage can be avoided in great measure through careful protection of the gonads during the X-ray examination. The frequency of spontaneous mutations in the mouse corresponds approximately to the mutation rates known for man. Among 429,240 young of female mice which were homozygous for seven known recessive genes and which were fertilized by homozygous normal male animals, there were 27 homozygous pathological animals. In these cases, then, a mutation must have occurred in one of the paternal germ cells. The mutation rate per locus found, by calculation from the seven loci investigated, is 0.9×10^{-5}. Among 48,500 young of male mice which had received a gonadal dose of 86 r in the form of chronic gamma rays, six mutations were observed, corresponding to a mutation rate of 1.77×10^{-5}. With a gamma irradiation of 37.5 r of cobalt-60 gamma rays, Carter and his collaborators obtained five mutations among 53,298 animals, corresponding to a mutation rate of 1.34×10^{-5}. Thus, a doubling of the mutation rate is achieved with doses of the order of magnitude of 40–80 r, which agrees approximately with the conditions for Drosophila.

The experiments with mice were carried out with animals which cannot be compared in a simple way with a human population. They involved matings of homozygous normal animals, in which the effect of the rays on the germ cells was to be tested, with animals which were homozygously abnormal for seven genes and whose germ cells always contained the pathological genes, and which therefore were suited for tests for the presence of recessive genes in the germ cells of the partner. Some pathological recessive genes are present also in human populations, but presumably there are fewer, even though probably not many fewer than seven per individual; but here they usually are present not in the homozygous but in the heterozygous state. Besides, entirely different types of genes are involved, whose homozygous effect often might be hardly identifiable in detail. Some perhaps manifest themselves in abortions or stillbirths, others in malformations, still others in recessive diseases which are not yet recognizable at birth or at the age of nine months. Therefore it should not be expected that the genetic damage in man from a single exposure to 300 r will be detectable in the children. Among 40,408 mice conceived by animals which had been exposed to an acute roentgen irradiation of 300 r, twenty-five recognizable mutations were found (mutation rate 8.85×10^{-5}), that is, one mutation for every 1,610 animals. As we have seen, the study by Neel and Schull concerning the children of the survivors of the atomic bombs of Hiroshima and Nagasaki who had received a comparable radiation dose included a total of 2,466 children. If the radiation-induced mutation rate in man were equal in magnitude to that of the mouse, and if existing conditions made it possible to demonstrate genetic damage in man as well as it is possible in the mouse, then one or two new mutations might have been found in these children. In a group of 26,369 children from parents at least one of whom had received 5–10 r no demonstrable damage would be expected, if the conditions regarding the radiation dependence of the mutations and the possibility of discovering them were the same for man as for the experimental animals. Finally, 6,278 children came from parents who had received an estimated 50–100 r. In this group, also, one could not expect a demonstrable amount of damage.

The radioactivity of the natural environment of man is estimated to be between 3 r and 4 r in the course of one generation. It is assumed that only a small part of the "spontaneously" occurring mutations may be explained by natural radioactivity. In our modern civilization, there is, besides the natural base radioactivity, an additional dose resulting from diagnostic X-ray examinations, which amounts perhaps to 1–3 r, that is, which can double the total dose. Compared to the amount coming from the natural environment and from X rays, the contributions of all further sources of radioactivity are small. For example, the radiation from luminous dials of wrist watches and from the fallout of the atom bomb tests carried out so far is estimated at 3–4 per cent, each, of the natural radiation load.

At this point, some comments on the etiology of leukemia seem appropriate. The cause of leukemia, as of other tumors, can be found in a somatic mutation. Ionizing radiations can cause leukemia. Among 95,819 survivors of Hiroshima, up to 1957, 68 had become ill with leukemia, or 0.071 per cent. Of 1,241 persons who were at a distance of less than 1 km from the central point below the bomb at the time of the explosion, 15, or 1.17 per cent, became ill. Of 1,722 children in whom hyperplasia of the thymus gland was "treated" by X-ray therapy, 7 (0.25 per cent) got leukemia. This is the more regrettable since the so-called hyperplasia of the thymus gland is almost always a completely harmless condition which is present in at least 30 per cent of all healthy infants. Of 13,352 patients with Bechterew's disease who were treated with X rays, 49, or 0.37 per cent, died of leukemia. In the group with the highest spinal marrow dose, 2,000 r or more, 2.1 per cent died of leukemia. Formerly, radiologists died of leukemia about ten times as frequently as did other physicians; according to an American statistical study, the figure was 4.7 per cent of all radiologists. Thus, leukemia is a real risk for the individual receiving large amounts of radiation. It is still unknown, however, what the danger due to smaller doses is. Data obtained so far can be brought into agreement with the assumption that the frequency of leukemia depends linearly on the total dose and that there is no threshold value below which a quantity of radiation is definitely harmless. If this assumption is true, 300 r could evoke as many cases of death from leukemia among 10,000 persons as 3 r among 1,000,000 persons. Under this presupposition, also, the fallout from the atomic bomb tests would lead to numerous cases of leukemia. It has been calculated that 25,000 to 150,000 cases of leukemia are to be expected as a result of the test explosions which occurred up to 1958. But since these cases are distributed over the billion population of the earth and also over the coming generations, and since they would represent only a minute fraction of all cases of death from leukemia, it will never be possible to test this terrible reckoning. These cases of leukemia, which in principle could have been avoided, compared to other cases where death could have been avoided in principle—perhaps those due to cigarette smoking and to traffic accidents— hardly carry any weight. However, this does not make them less tragic. The physician does not think of the mortality statistics so much as of the individual. He should always keep in mind that, in principle, every radiograph can lead to the death of a person from leukemia. Measures such as series of X-ray examinations of an entire population should be carefully weighed for their potential usefulness and damage. In an English report for the year 1957, an attempt was made to set up a quantitative balance sheet. During that year, by means of radiographic screening pictures, 17,835 cases of tuberculosis, 12,000 cases of heart disease, 9,400 cases of silicosis of the lungs, and 2,362 cases of lung cancer were discovered. Under the most unfavorable assumptions, 20 cases of leukemia might have been expected as a result of these X-ray examinations.

The quantitative difference between mutations in the germ cells and the somatic mutation which forms the basis of leukemia probably is due to the fact that only mutations in those germ cells which happen to undergo fertilization can lead to a recognizable effect. In every ejaculate, which contains several hundred million spermatozoa, there are, on the average, many thousand pathologically mutant genes which can lead to a hereditary anomaly in the child. The chance that a spermatozoon with a mutant gene will win the race is nevertheless relatively small. When a few hundred genes which have undergone mutation as a result of radioactivity are added to the thousands of pathological genes in an ejaculate, it is hardly possible to demonstrate an effect. With the somatic mutation which is the basis of leukemia, however, the situation is different. It is estimated that 9×10^{12} leukocytes and lymphocytes are present in the entire body. If approximately every thousandth cell among these is still young enough to turn from a leukoblast into a leukemia cell, there would be 9 billion potential leukemia cells. If we assume that the chance for each individual cell of experiencing a mutation leading to leukemia because of a given dose of radiation is only 1 per 900 billion, such a radiation dose would be capable of producing one case of leukemia among 100 individuals. We would, however, never be able to detect a radiation dose which produces one mutation per 900 billion germ cells. The order of magnitude of detectability is basically different for genetic and for somatic mutations which can lead to malignant tumors. For the geneticist, who has in mind the genetic fate of the population, leukemia represents a danger signal which beams a warning before the genetic danger of the radiation exposure becomes recognizable. It is to be hoped that this signal will be heeded earlier than the geneticist's warning of a danger which cannot be directly seen or measured.

COMPOSITE GENE EFFECTS

Interaction of non-allelic genes

If a trait or a disease depends on a single gene or pair of alleles, we speak of monomeric or monogenic inheritance. Up to now, we have discussed only such monomeric normal and pathological traits. However, it is also conceivable that a disease may come about only when two different genes meet which are not detectable individually. In man, this possibility has not been demonstrated.

First, we shall be concerned with the interplay of four non-allelic pairs of genes in the domain of blood groups, namely, of the ABO blood groups with the Lewis factor, the secretor factor, and the xx-suppressor gene. The property of the red blood corpuscles of reacting with anti-A or anti-B serum or both depends, for practical purposes, exclusively on the particular genes of the ABO allele series which the individual in question possesses. Some persons have these antigenic properties not only in their blood corpuscles but also in some body fluids, especially in the saliva. Such persons are called secretors. A dominant gene, Se, is responsible for the secretor characteristic. Persons of genotype SeSe or Sese secrete A as a blood group substance, if they also have the gene for blood group A. In contrast, in the saliva of a person with the blood group gene formula AA, AB, or AO, no A substance is detectable if he is a non-secretor, that is, if he is homozygous for the recessive gene se. Thus the property of "blood group substance A in the saliva" is dependent on two gene loci. The gene pair SeSe, Sese, or sese has a still deeper significance for the manifestation of another pair of genes, namely, the pair of alleles L-l (Lewis factor). This pair of alleles is responsible for the reaction of saliva and of the red blood corpuscles with anti-Le^a. (The genes of the Lewis system are designated by L and l, and the corresponding antigen by Le^a.) The Lewis substance is not primarily bound to the red blood corpuscles, but is only secondarily adsorbed by these. Thus, in a blood transfusion, the erythrocytes of a donor who was originally Le^a-negative can take on the Le^a substance from the plasma of an Le^a-positive recipient and become Le^a+. In secretors, the Lewis substance is never found in the saliva or in the erythrocytes, but in non-secretors almost always. These peculiar relationships seem to be due to the fact that the same basic substance is used for the ABO blood group substance and for the Lewis substance. If a person is a secretor, the entire basic substance is utilized for the production of the ABH substance, and not enough is left for the Lewis substance.

It has been possible to show that the ABH properties and the Lewis[a] properties are located in the same molecule and that they are closely related chemically. The substances A, B, and Le[a] contain the same four sugar components: l-fucose, d-galactose, d-glucosamine, and d-galactosamine, and the same eleven amino acids. According to investigations with the ultracentrifuge, electrophoresis, and fractionated precipitation, they are very similar. Their molecules are of the same size of $200–300 \times 10^3$, and of equal asymmetry. Finally, there are antisera which react only with certain configurations of A and L alleles, such as the so-called anti-Le[b]. But the property Le[b] is determined not by a monomer, but by a specific combination of alleles of the ABO series, the Sese series, and the Ll series. Magard serum reacts similarly with a factor originally called "Magard" which is determined by the combined effect of at least one A gene, one Se gene, and the absence of the L gene. Thus, there are rather good reasons for the assumption that the blood group substances A, B, and Le[a] are different forms of a basic substance, where this basic substance had imprinted on it by the different non-allelic pairs of genes the specific antigenic properties. Fitting this concept is the fact already mentioned, that the antigenic properties A and B are not distributed over two different types of molecules, in individuals who have the two responsible genes simultaneously, but that they are located in the same molecules. One would assume that the antigenic properties A and B must be tied to different molecules, if the genes for A and B produced the antigenic substances as direct gene products. The basic substance on which the antigenic properties AB can be imprinted by the specific blood group genes is presumably supplied by a third, non-allelic gene which has been designated as X. This assumption is supported by several observations in which the erythrocytes and the serum contained no A or B antigens, so that the blood group determination at first led to the assumption of blood group O. However, this was wrong because of the absence of the antigenic property H in the erythrocytes (phenotype O_h), and the presence of a strong anti-H in the serum, which reacted with all erythrocytes of persons in blood group O. In general, all persons in blood groups O and B also have the antigenic property H in their erythrocytes and, if they are secretors at the same time, in their saliva as well.

Another argument against a genotypic O/O constitution is the fact that such atypical pseudo-O mothers can have children in blood group A_1B. Thus, the mothers must possess one of the two genes—in the published cases, the gene for B. One assumes that, in these women who apparently defy the Mendelian laws, a recessive suppressor gene x is present whose homozygous state makes the formation of B and H blood group substance impossible. Of course, the characteristic of the recessive suppressor gene x/x cannot be determined directly, but only through the irregularities of the phenotypic manifestation of other blood group genes, which are due to its presence. The assumption that a recessive gene is involved is based on observations of phenotype O_h in siblings and in children of consanguineous parents. It would be conceivable that the H property is assigned to the normal dominant allele X of the suppressor gene x. Since individuals without ABH antigens cannot be secretors for these antigens, the manifestation of the Se gene is also impossible in the presence of the homozygous suppressor gene xx. Here again, however, the suppressed dominant gene Se can reappear in the children, when the homozygous x/x genes have again been separated. In spite of the presence of Se, it finally becomes possible, contrary to the rule, to detect the Le^a factor through a homozygous x/x, since now the basic substance for the formation of the Le^a factor is not used up by the competition of genes A and B. This is a further confirmation of the fact that the Se

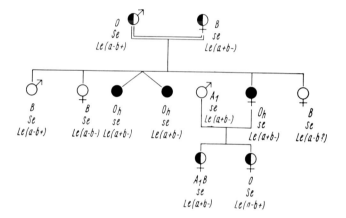

Fig. 58.—Blood group phenotypes in a family in which II_3, II_4, and II_6 are homozygous for the recessive suppressor gene x.

gene has no direct connection with the Le^a factor, but that its effect on the basic substance merely inhibits the manifestation of the Le^a factor. In this way, a single homozygous pair of genes can produce the abnormal phenotype O_h from the genotype B/O, the abnormal non-secretor phenotype from the genotype Se/se, and, from the genotype LL Se/se, the phenotype $Le^a +$ which is abnormal for this configuration.

Under normal circumstances, the presence of the isoagglutinins anti-A and anti-B depends solely on the ABO genotype of the individual in question. Individuals in blood group A form anti-B, while individuals in blood group B form anti-A. Individuals in blood group AB form neither anti-A nor anti-B. In the same manner, persons who have no H antigen form anti-H. Persons known up to now who are of genotype xx were first recognized by the presence of a strong anti-H. H-, A-, and B-negative xx-individuals will also form anti-B if they have genotype BO, such as the female patient II_6 in Figure 58. The formation of the naturally occurring isoagglutinins is thus dependent not on the genotype directly, but on the gene product.

There is another interesting exception to the rule according to which persons always form antibodies against those ABO blood groups which they do not possess themselves. This exception, which we can also regard from the point of view of the interaction of non-allelic genes, is X-chromosomal recessive agammaglobulinemia. In this anomaly, γ-globulin is missing. Since the isoagglutinins belong to the γ-globulin fraction, patients with agammaglobulinemia, even if they have blood group O, cannot form any anti-A or anti-B. The material for the formation of antibodies is not available. The normal allele of the X-chromosomal gene for agammaglobulinemia is evidently one of the prerequisites for any formation of isoagglutinins.

From pathology, only a few examples of the interaction of non-allelic genes are known. In general, the result of the getting-together of several non-allelic pathological genes is simply an additive combination of two diseases and not

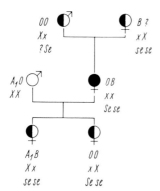

FIG. 59.—Accompanying genotypes, to the extent that they can be inferred from the phenotypes and from the relationships between family members.

a basically new disease pattern. Thus, the combination of spherocytic hemolytic anemia and sickle cell anemia is purely additive. If some diseases occur in combination with greater than chance frequency, as, for instance, obesity and diabetes mellitus, these are almost always diseases which do not depend in their etiology simply on monomeric genes, but on multiple genetic and exogenous factors. Here, one disease can become a pathogenic factor in the complex mechanism of the other disease. The significance of adiposity, or the overeating correlated with it, for diabetes mellitus, or the significance of diabetes mellitus for atherosclerosis can probably be interpreted in this sense. Correlations between diseases thus are not based on connections within the genotype but on common factors in the pathogenic mechanism of diseases determined by a multiplicity of factors. In the genotype, the pathological genes are not "co-ordinated," as one sometimes reads in older publications. Only their phenic effects are sometimes co-ordinated.

One of the few clear examples of the interaction of non-allelic genes in the etiology of a disease is provided by the combination of the heterozygous genes for thalassemia and sickle cell anemia. Presumably, the genes for thalassemia and for sickle cell anemia are not alleles. If we call the gene for thalassemia th^T, the normal allele th^A, the gene for sickle cell hemoglobin, Hb$^S_\beta$, and the normal allele for it Hb$^A_\beta$, the following combinations are known:

Hb$^A_\beta$Hb$^A_\beta$	th^{A}th^A	normal
Hb$^S_\beta$Hb$^A_\beta$	th^{A}th^A	sickle cell feature
Hb$^S_\beta$Hb$^S_\beta$	th^{A}th^A	sickle cell anemia
Hb$^A_\beta$Hb$^A_\beta$	th^{T}th^A	thalassemia minor
Hb$^A_\beta$Hb$^A_\beta$	th^{T}th^T	thalassemia major
Hb$^S_\beta$Hb$^A_\beta$	th^{T}th^A	sickle cell thalassemia (double heterozygotes)

(Combinations of homozygous th^T with Hb$^S_\beta$, or of homozygous Hb$^S_\beta$ with th^T, are still unknown.)

The interaction of the two pathological genes is shown in the fact that the double heterozygotes have a considerably more severe anemia than the two simple heterozygotes. The hemoglobin level lies between 6 and 10 g per cent; at the same time, microcytosis and sickle cell formation are present in vivo. The hemoglobin consists up to 80 per cent of hemoglobin S; fetal hemoglobin can amount up to 17 per cent. In some cases, the production of hemoglobin A is so severely inhibited that no distinction is possible from the hemoglobin (S + F) of homozygous sickle cell anemia. Only examination of the family can then show whether the individual is doubly heterozygous or homozygous for Hb$^S_\beta$. Clinically, double heterozygotes can exhibit severe anemia with enlargement of the spleen, abdominal crises, ulcers of the legs, and crisis-like pains in the joints which are reminiscent of acute rheumatism. All these pathological symptoms are absent in thalassemia minor as well as in the sickle cell trait. A similar interaction appears to exist between the gene for thalassemia and the gene for hemoglobin H; in fact, hemoglobin H, which consists of four β chains, becomes detectable only if a gene for thalassemia is present in addition to the specific gene for this pathological hemoglobin.

A somewhat more complicated example is that of fetal erythroblastosis due to Rh antibodies which are transmitted from the Rh-negative mother who is sensitized against Rh, to the Rh-positive child, and which destroy its erythrocytes. Here, also, a series of alleles which is independent of the Rh system is involved, namely, that for the ABO system. The sensitization of the mother against the Rh factor is brought about by fetal erythrocytes which enter the maternal circulation. One would assume that the sensitization of the mother against the Rh factor should be independent of the ABO blood groups of mother and child. In actual fact, however, sensitization against Rh sets in much more readily if mother and child have the same ABO group than if the mother possesses isoagglutinins against the child's erythrocytes. The cause of this involvement of the ABO system in Rh-erythroblastosis probably lies in the fact that, in the ABO system, erythrocytes of the child which are in the same blood group remain in the maternal circulation considerably longer, so that their Rh property is able to exert a stronger effect on the mother. The role of ABO compatibility therefore extends only to the immunization of the mother, which usually sets in during the first pregnancy. On the other hand, the ABO type of subsequent children is immaterial for their fate, since the mother now already possesses Rh antibodies.

An interaction of non-allelic genes probably takes place also in the relationships between the frequency of ulcus duodeni and the ABO blood groups as well as the secretor property. Among patients with duodenal ulcers, blood group O occurs approximately 1.2 to 1.6 times as frequently as among random persons, and the secretor characteristic is twice as frequent with them. In spite of the confirmed statistical relationship, however, a plausible explanation is lacking, so that this example hardly serves to advance our understanding of gene interaction. It shows, however, that one must consider an interaction on occasion, even when one would hardly suspect it. It may probably be assumed that the blood group substances of the ABO system have important functions which are still unknown and that they do not exist merely for the purpose of making blood transfusions complicated. According to Vogel's concepts, the decisive biological significance of the ABO blood groups is to be looked for in antigenic properties which they have in common with the causative agents of the great epidemic diseases (plague, smallpox, syphilis).

Sex-limited or sex-controlled manifestation

Genes located in the X chromosome are called sex-linked. Sex-linked genes can become effective in the phenotype in both sexes. Their relation to sex is completely explained by their dominant or recessive effect and by the chromosomal mechanism of sex determination. The term sex-limited, on the other hand, is used for those anomalies in which the responsible genes are inherited independent of sex but become effective exclusively or preferentially in one of the two sexes. Sex-limited anomalies therefore cannot be called dominant or recessive in the simple sense, since the effect of the responsible gene does not depend only on the one pair of alleles but also on the sex of the carrier of the gene.

There are only two examples of the sex-limited manifestation of genes in man which are genetically transparent to some degree: pubertas praecox of boys, which is dominantly inherited, and so-called testicular feminization. Pubertas praecox is the abnormally early occurrence of an otherwise normal development of puberty. Pubertas praecox occurs more frequently in girls. However, the causes of pubertas praecox in the female sex are known only for a small portion of cases, in which lesions of the hypothalamic centers due to inflammatory changes, hydrocephalus, or tumors can be held responsible. There is a tendency to designate the remaining cases as "constitutional" pubertas praecox, although the word "constitutional," here and in most other instances, does not have a very clear meaning. In contrast, pubertas praecox in boys, though often also due to localized disorders of the midbrain, exists in addition in a dominant hereditary male form. In such boys, the first signs of puberty appear during the second year of life. At four years, the development of the genitals and the growth of secondary hair usually correspond to those of an adult man. These precocious children are also far ahead of their age in body growth, but not in their mental development. The anomaly can be transmitted to male descendants by healthy women whose sexual development begins at the usual time. To this extent, the pedigree is similar to that of a recessive sex-linked hereditary disease. Indeed, several pedigrees have been published which do not deviate in any way from the conditions existing in X-chromosomal hereditary transmission. In three of these pedigrees with a total of seventeen afflicted male family members, however, no afflicted man had a son. Thus it is impossible to decide whether the anomaly is an X-chromosomal recessive one or a dominant one with a sex-limited manifestation. The decision can be made only if the

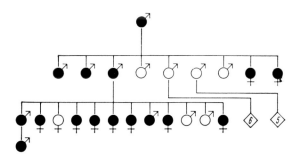

FIG. 60.—Autosomal dominant inheritance of sterility with restriction to the female sex. Syndrome of peripheral ptosis, blepharophimosis, and epicanthus. (Pedigree from Usher.) In typical cases, the female patients usually have no children.

afflicted men have male descendants. In the two additional pedigrees where this was the case, five afflicted fathers had a total of ten sons, six of whom were afflicted and four normal. In X-chromosomal inheritance, transmission from father to son would be impossible.

The second example of a dominant anomaly with sex-limited manifestation is so-called testicular feminization. Testicular feminization refers to a peculiar intersexual condition in which the patients, who are genetically male, that is, who have an XY set of chromosomes, have internal testicles but female external genitals and female breasts. This condition is transmitted by healthy women who, however, just as the patients themselves, have only sparse axillary and pubic hair, or none at all. The mechanism of the condition is presumably a lack of response of the cells to androgenic hormones. Androgenic hormones are responsible for the development of the male genitals. If the organs do not respond to the androgens, the genitals remain in their original neutral form, which corresponds to the form of the normal female sex. In the genetically female carrier of the trait, the missing reaction to the androgenic hormones shows itself only in the non-appearance of secondary hair growth. The basic defect, namely, the resistance to androgens, is a dominant hereditary trait. The basic defect is of pathological significance only in the male sex, however. In this respect, the syndrome is sex-limited. Testicular feminization can be designated as a dominant hereditary disease with manifestation limited to the male sex also for the simple reason that, in most cases, no information is available about the harmless absence of hair in the female gene carriers. Intersexual patients are, of course, sterile, so that they cannot transmit the gene. Thus, without studies of linkage, a definite decision is impossible concerning whether this is a dominant anomaly with limitation to the male sex, or a recessive X-chromosomal anomaly.

This restriction must be kept in mind for all presumably X-chromosomal recessive anomalies in which the patients are incapable of propagation or die

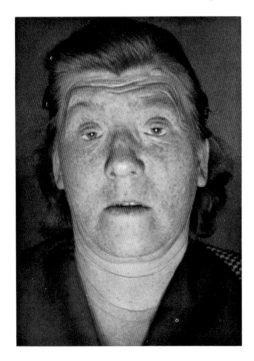

Fig. 61.—Ptosis with blepharophimosis and epicanthus. Anomaly with sterility limited to the female sex.

before reaching the age of reproduction. Only through the sons of male patients can it be determined with certainty whether a disease is sex-linked or sex-limited. It can be assumed in general that a sex-limited manifestation occurs only in functions or organs which normally exhibit distinct sexual differences. Anomalies of the skin, the eyes, the nervous system, or the internal organs, which are transmitted only by healthy women, can be regarded as being almost certainly of the X-chromosomal recessive type.

Most of the sex-dependent anomalies, such as Bechterew's disease, Dupuytren's contracture of the palmar aponeurosis, ulcus duodeni, Basedow's disease, or congenital luxation of the hip, have, on the basis of all available information, no simple etiology which can be explained by a single gene, but are determined by several factors, some hereditary and some non-hereditary. In conditions that can be affected by many factors and thus are probably more labile in their development, it is reasonable to expect that factors related to sex, especially those of a hormonal nature, can exert an influence. However, if simple dominant anomalies, such as hereditary drumstick fingers or multiple exostoses, are sex-controlled in their manifestation, the difference between the sexes is found on closer observation rather to be one of degree. Corresponding to the greater growth of bones and extremities of the male under hormonal influence, cartilaginous exostoses and drumstick fingers here also develop more strongly. In women, these anomalies escape diagnosis more readily, since they are less clearly defined. It is further characteristic of this type of sex-dependent manifestation that the sex difference appears only at puberty, at which the manifestations develop more clearly in boys.

It must be emphasized, however, that in many cases no plausible explanation is yet known for the differential frequency of diseases in the two sexes as, for example, for the distinct predominance of boys in the case of spastic hypertrophy of the pylorus of infants, or of girls in the case of anencephaly.

Sometimes only one symptom of a disease is sex-limited. Thus it appears that a dominant hereditary anomaly of the eyelids which is designated as (peripheral) ptosis with blepharophimosis and epicanthus causes sterility only in women. The pedigrees show that the anomaly is transmitted only by afflicted men, while afflicted women usually have no children (Cockayne, Edmund, McIlroy, Usher, Vignes, Waardenburg). In one family, which I was able to observe with Nowakowski, the proband and two afflicted sisters of her father, who was also afflicted, had had childless marriages of many years' duration. There was a secondary amenorrhea with hypoplasia of the ovaries. Ptosis with blepharophimosis appears to be an autosomal dominant hereditary disease with complete sex limitation of one symptom, namely, that of sterility.[1]

Irregularities of gene effects. Penetrance and expressivity

Up to now we have discussed almost exclusively genes whose phenotypic effect depends in a known and unique manner on the genotype. The simplest examples were those anomalies in which the heterozygous or the homozygous state of certain genes is always recognizable by a certain phenic manifestation. Besides, we have already come to know examples in which the manifestation of a gene depends on the simultaneous presence of other allelic or non-allelic genes. Such an example was the manifestation of the gene for blood group property B, which depends on the presence of the dominant factor X. If we now imagine that the recessive factor x is relatively

---

[1] Owens, Haley, and Kloepfer recently found in a larger compilation seven female patients who had children, as compared with twenty-four male ones. Thus, the sterility does not seem to be absolute. The reverse possibility, i.e., sterility limited to the male sex, might offer an explanation for the curious fact that pseudohypoparathyroidism is transmitted almost exclusively by females.

frequent in the population, so that it would occur in the homozygous state fairly often and could thus suppress the manifestation of the gene for B, and if we had only an anti-B serum but no anti-H, then we conclude that the antigenic property B of the erythrocytes has irregular dominance. In a certain percentage of cases, let us say 5 per cent, it would not be detectable in spite of the presence of the corresponding gene. We would then say that the gene for blood group property B has a penetrance of 95 per cent. Actually it is customary to reserve the expression "penetrance" for cases in which the cause of the missing manifestation is unknown. The causes which stand in the way of 100 per cent penetrance do not always have to be genetic ones. Let us assume that the gene for the lack of glucose-6-phosphate dehydrogenase of the erythrocytes is relatively rare in a population, but that this population lives on a diet a regular constituent of which is vicia fava, and that only a few individuals in it, for some reason, eat no vicia fava. If the familial character of favism were known while its relationship to the ingestion of vicia fava were still concealed, we could again speak here of a gene with incomplete penetrance; but in this case the cause of incomplete penetrance would be found in the diet, that is, in exogenous factors. Here, also, we would no longer speak of penetrance as soon as the true relationship became known.

Penetrance is a concept which attempts to bridge the discrepancy between theoretical expectation on the basis of a genetic hypothesis, and actual observation. The concept of penetrance is by its nature a temporary, purely formal auxiliary hypothesis without specific content, which is intended to make a genetic hypothesis acceptable. If the discrepancy between observation and expectation is considerable and its cause unknown, doubts about the hypothesis which was made the basis for the theoretical expectation are indicated. In some cases, however, the concept of penetrance has a more exact meaning, namely, for pathological traits whose regularly dominant inheritance in extensive kindreds is immediately recognizable, but in which individual members are occasionally skipped. Such a condition is found, for example, in the case of familial pubertas praecox of boys. Ordinarily, the male individuals who possess the responsible gene exhibit typical pubertas praecox. It happens in individual cases, however, that sons of precocious fathers reach puberty at the normal time themselves, but that they again have precocious sons. One reads occasionally that it happens very often in man, with the dominant mode of inheritance, that a generation is skipped; this seems to me to be based on a too generous application of the concept of dominance. There are numerous anomalies with irregular familial occurrence whose genetic nature has not yet been explained in detail. Among such anomalies, usually the "interesting" pedigrees whose many affected members can evoke the impression of dominant inheritance are published first in the case literature. Also, some pedigrees with "irregular dominance" are published, and one then speaks of "missing penetrance." More extensive statistical studies, which start with probands who were not selected because of familial occurrence of the anomaly, show, however, that no simple gene hypothesis gives satisfactory agreement with the observations. In such cases it seems more proper to me to admit our ignorance openly than to adapt a scheme of dominant inheritance to the observed figures with the aid of the elastic hypothesis of penetrance.

Penetrance refers to the presence or absence of the phenic trait for a given gene. Differences in the definition of the trait due to the gene, on the other hand, are called differences in expressivity. The concepts of expressivity and penetrance can overlap to the extent that the weak definition of a trait (low expressivity) may be overlooked, so that the impression of missing penetrance results. The more accurate our methods of investigation, the more certainly we shall observe even small deviations, and the less frequently we shall have to resort to the concept of missing penetrance. Hemochromatosis serves as an example of this. If we have in mind only the fully defined disease syndrome of bronze diabetes with skin pigmentation, cirrhosis of the liver, and diabetes mellitus, then the responsible dominant gene has irregular penetrance in the male and very weak penetrance in the female sex. If we study the serum iron level, penetrance becomes complete in the male sex after puberty, but remains low in the female sex. But if we study the iron uptake of the intestines, we regularly find a pathological increase in both sexes, that is, 100 per cent penetrance for the metabolic peculiarity which is most closely related, according to our present knowledge, to the primary gene effect. In the female sex, the pathological consequences of increased iron uptake are usually prevented by the regular menstrual blood loss. It is of great theoretical and practical interest that this cause of the missing manifestation of the complete disease syndrome can be induced artificially in the male sex, namely, by repeated venesections. Patients having hemochromatosis should be lifelong blood donors, for they are then protected to a large extent from the pathological deposit of iron-containing pigments in liver, pancreas, and spleen, which leads to the severe pathological syndrome.

In any case, penetrance and expressivity are not properties of the genes in question, but functions of the rest of the genotype and of the environment. The primary gene effect usually depends only on a single pair of alleles, whereas numerous genetic and exogenous factors are combined in secondary and tertiary effects. The concepts of penetrance and expressivity belong in the realm of these secondary and tertiary complex gene effects, but not in the realm of monomeric gene action. Thus, if it is said of a gene that it is dominant and has a penetrance of 50 per cent, this means that the effect of the gene depends on additional exogenous or genetic factors which are still unknown. The concept of penetrance therefore does not solve a problem but rather poses it. Sometimes one has the impression that the concept of penetrance serves in the literature to obscure a problem. As soon as one compiles anomalies with a similar clinical picture from different kindreds, justifying the concepts of penetrance and expressivity becomes very questionable, since one no longer knows whether the same gene is responsible. In the many hereditary anomalies which are essentially determined by a single pair of alleles, the penetrance is usually 100 per cent, but the expressivity shows smaller fluctuations the more we succeed in recognizing the primary gene effect and taking it to be the decisive feature of the anomaly. For all hereditary anomalies with defective penetrance, that is, a penetrance of less than 90 per cent, it is advisable not to speak of simple dominance or recessivity. Each anomaly with low penetrance or with strong fluctuations of expressivity confronts the research worker with the task of uncovering the causes of the absent or variable manifestation of the genes. The explanation of the causes of defective gene manifestation promises to provide important information for the prophylaxis and therapy of genetically determined disorders.

Gene and environment

Properties which clearly depend on individual genes seldom show an equally clear capacity for being influenced by environmental factors. Nevertheless there are such examples. Galactosemia is a recessive hereditary anomaly in which the body cannot utilize galactose in the normal way. Under normal environmental conditions, that is, with breast feeding, the infant always obtains galactose through the milk sugar (1 molecule of milk sugar is composed of 1 molecule of galactose and 1 molecule of glucose). A child receiving a normal diet, but having galactosemia, therefore becomes ill with cirrhosis of the liver, cataracts, and feeble-mindedness. It is possible, however, to eliminate galactose from the diet. Such children can then develop normally in spite of their metabolic anomaly. In the case of fructose intolerance, which is also recessively inherited, the administration of fruit sugar or of beet sugar (which is composed of 1 molecule of fruit sugar and 1 molecule of glucose) leads to severe hypoglycemic attacks with nausea, bloody vomiting, tremors, perspiration, and somnolence. All these symptoms are absent if no fructose is given to the children. In alactasia, the enzyme which breaks down the milk sugar in the intestine is missing. Infants have fermented stools and do not thrive because the normal source of carbohydrates from the breast milk diet is eliminated. If they are given another sugar, no further disorders occur. The symptoms of essential hyperlipemia, which consist of crisis-like severe stomach pains with spleen and liver enlargement, and in older patients also of angina pectoris, can largely be avoided by means of a low-fat diet. A special genetically determined abnormality of the enzyme pseudocholinesterase might not have been discovered if it were not for the fact that, during the past few years, succinyl choline chloride has proved itself a reliable short-term muscle relaxant. It soon became apparent that, in rare cases—about once among 5,000 patients—succinyl choline chloride leads to prolonged cessation of respiration. Accurate examination of patients who had experienced such anesthesia incidents uncovered an abnormal pseudocholinesterase level in their blood. It is not known whether this enzyme defect has any pathological significance at all, apart from the abnormal reaction to succinyl choline chloride.

In these examples, the relationship between a specific gene effect and an exogenous substance is quite clear. It is to be assumed that numerous other cases of this type will still be discovered, especially in regard to intolerance to drugs. Here, we shall first be further concerned only with a quantitative consideration. If, in the etiology of a pathological condition, two factors are involved side by side, of which each alone does not lead to the pathological condition, that factor which is most rare is the more important one etiologically, while a very frequent factor is of little significance etiologically. This may seem surprising at first because it goes against our sensibility to designate a frequent factor as insignificant and a very rare one as important.

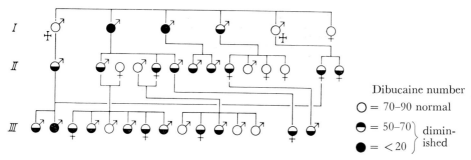

Fig. 62.—Familial pesudocholinesterase abnormality. The pseudocholinesterase activity was measured by an inhibition test with dibucaine. For I_1 and I_6, no determination was made. Several family members who had already died and had not been examined were omitted. It is characteristic for a recessive feature (with partial manifestation of the heterozygous state) that the two homozygotes I_2 and I_3 had only heterozygous children, and that the homozygote III_2 came from a cousin marriage. (From Kalow and Staron.)

Yet this connection between frequency and relative significance of etiological factors is firmly rooted in general usage and in our thinking; but we are usually not aware of this. To continue with our last example, if not 0.02 per cent, but 90 per cent of the population suffered from pseudocholinesterase deficiency, we would no longer have any reason for observing a differential reaction to succinyl choline chloride. We would regard the substance as a severe poison without inquiring much into the different reactions to it. The problem is similar in psychogenic illnesses. Even if it could be demonstrated by a flawless method that specific psychically effective configurations are always responsible for a specific set of symptoms, perhaps bed-wetting or asthma, the fact would still remain that the same configurations are so frequent that one cannot ascribe a decisive meaning to them in the sense of our etiological scheme. The quarrel over the relative contributions of heredity and environment here often develops into a fruitless argument over concepts. For those who regard the patient as possessing given individual characteristics, the decisive question is what environmental influence has produced the symptom in this patient. On the other hand, someone who is interested in the general question of etiology will easily arrive at the point of regarding psychological difficulties as everyday, general events in this world but will go in detail into the question why just this specific patient cannot cope with the difficulties. This second example, which is intentionally somewhat general, should by no means lead to the opinion that the methodology of human genetics offers patent solutions to complicated problems of this type. It can only help to clarify the complex questions posed, and point out that the individual is not only a product of experiences of his early childhood and of errors in his environment, but that he also carries within him his own response potential.

Phenocopies

If it is possible to evoke the same phenotypic anomaly either by a specific genotype or by an exogenous factor, the anomaly is referred to in the latter

TABLE 24

RELATIONSHIP BETWEEN TIME OF INFECTION
AND FORM OF MALFORMATIONS IN
RUBELLA EMBRYOPATHY

Infection during Weeks of Pregnancy	Malformations of the Embryo
4–5	Cataract
5–7	Cardiac defects
6–14 (8–12)	Damage of inner ear
8–9	Defects of deciduous teeth

case as a phenocopy. Thus, a phenocopy is an imitation, by non-genetic causes, of a condition which is usually genetically determined. Phenocopies in this sense are not known in man. However, if one is less demanding in his definitions and also uses the term "phenocopies" for those anomalies of an exogenous nature which exhibit a certain similarity with known, genetically determined defects or with defects in whose creation genetic factors at least participate, individual examples of phenocopies can be pointed out. If a mother becomes ill with rubella during the first third of her pregnancy, the embryo is damaged in about 20–30 per cent of all cases. The embryopathy of rubella has a characteristic symptomatology, which evidently depends primarily on the specific nature of the rubella virus, since no comparable consequences during pregnancy are known for any other virus infection. The development in an individual case, however, depends on the time of infection. Diminished growth, feeble-mindedness, microcephaly, clouding of lenses, heart defects, damage to the inner ear, and milk tooth defects are characteristic of the embryopathy; the relationship of these disorders to the time of infection appears in Table 24.

There are also cases of clouding of the lenses and disorders of the inner ear which are determined purely genetically, and to this extent the embryopathy of rubella could be called a phenocopy. But usually an unequivocal decision can be made whether one is dealing with an illness determined by heredity or by exogenous causes, without knowledge of the previous family history and of the pregnancy anamnesis, purely on a phenotypic basis by an accurate ophthalmological and general examination. Thus, the phenocopy is a bad copy of he model. Cases of accidental roentgen irradiation of the embryo in the mother's body can serve as a further example of a phenocopy in man. Here, also, a characterisic syndrome results which has cataracts, microcephaly, and inhibition of growth in common with the rubella embryopathy, but which ordinarily is not accompanied by malformations of the heart or of the inner ear. Instead, the growth of the long bones can be more strongly affected in the case of roentgen damage of the embryo.

On the basis of certain animal experiments, it has been concluded that most malformations owe their specific character not to the nature of the experimentally established damage but to the timing of its effectiveness. Explanations of many human malformations have subsequently also been attempted, essentially on the basis of the supposed time of causation, considering particularly exogenous influences such as virus infections, dietary insufficiency, or lack of oxygen. Apart from the embryopathy of rubella, however, no definite knowledge is available about whether malformations in man can be traced back to such factors. Experience with the embryopathy of rubella and with radiation damage of the embryo is contrary to the idea that one and the same type of damage can cause a great variety of malformations depending on the time of its effect. Most of the frequent malformations—spina bifida; anencephaly; cleft lips, jaws, and palates; malformations of the extremities; and malformations of the urogenital organs—do not seem to occur more frequently than expected in children with rubella or roentgen damage. For the majority of these malformations the cause is unexplained. If it could be demonstrated that a definite exogenous injury was responsible for a case of anencephaly or spina bifida, one could still not speak of a phenocopy, since the genetic nature of the prototype is questionable. Among the numerous, diverse typical anomalies with a known mode of inheritance—dysostosis cleidocranialis, the nail-patella syndrome, chondrodystrophy, osteogenesis imperfecta, neurofibromatosis, chondromatosis, multiple cartilaginous exostoses, brachydactylies, polydactylies, etc.—in no case has a confirmed exogenous origin, that is, a phenocopy, been observed.

The layman's demand for causality seeks an explanation for an unexpected event such as a miscarriage. The most diverse possibilities are readily regarded as realities in this situation. The gullibility of physicians in regard to the attempts at explanation popular during the past few years, such as vitamin deficiency, oxygen deficiency, or psychological effects, has hardly been less than that of the lay public, so that a large part of the literature about malformations should be read with critical caution. Anyone who is more deeply interested in the genetic aspect of the problem is referred to the expositions by Neel and by Fraser (1959). Today, the boundaries of our lack of knowledge concerning the etiology of congenital malformations can be staked out only approximately: The percentage of malformations with known genetic etiology certainly is far below 10 per cent of all malformations; the proportion of malformations with known exogenous etiology is negligibly small and probably lies below 1 per cent. It is not to be expected that the remaining more than 90 per cent can be traced back to simple etiological factors. The assertion that in most cases of malformations heredity and environment are both involved is also not based on factual evidence. At the moment, any more definite judgments about the participation of genetic or exogenous factors in the etiology of the major proportion of human malformations would be premature. It is true that conclusions based on analogies with malformations in animals which are determined by heredity or produced experimentally by exogenous factors can show us what possibilities must be considered in man, but they can never enable us to discover what factors are actually in effect in man. Conclusions by analogy from animal experiments have a certain value in the formation of hypotheses, which must be the basis of all fruitful research; but the testing of the hypotheses is possible only by means of studies on man. Some schematic representations which have been used in many textbooks are misleading, namely, those where the etiology of malformations is shown as a circular area whose sectors are occupied by the individual, presumably effective, etiological factors and where only a small sector is set aside for unknown causes, although numerous factors are cited whose significance for human malformations has not been proved. We know from animal experiments that the frequency of some malformations increases under certain experimental conditions, such as oxygen deficiency or cortisone therapy during pregnancy; but we also know that the genotype of the strain of animals under investigation is of decisive significance for the frequency of exogenously produced malformations. It will be necessary to consider a similar interaction of the genotype and exogenous factors in man.

Polygenic inheritance (Polymerism)

GENETIC DIFFERENCES DETERMINED BY NUMEROUS GENES

So far, chiefly those genetic properties have been discussed which make it possible to divide the population into two or three genotypes, namely, individuals having the trait in question, individuals not having the trait, and sometimes still a third group of individuals in whom the trait is only weakly expressed. Most pathological genes, but also a considerable number of normal genes such as those for the blood groups, permit such a clear-cut bi- or trisection of the population. In addition to these traits, whose behavior usually agrees with simple hypotheses such as recessive or dominant, autosomal or X-chromosomal mode of inheritance, there are many important heredity-determined differences between individuals which do not show a sharp bi- or trisection within the population but rather a continuous variability. A genetic analysis of such quantitative, continuously variable features is possible only in animal experiments, where hybridization and breeding can be carried out at will. We know from experimental genetics that the quantitative variability of most normal morphological and functional properties is based on the interaction of a multiplicity of individual genes, no single one of which exerts such a strong effect that carriers of the gene can be distinguished with certainty from individuals having other alleles. The assumption of the genetic nature of the normal variability in man is based, first, on the proof of a resemblance between blood relatives and, second, on the proof that environmental factors exert no noteworthy influence. Research on twins has proved to be especially valuable in the study of the hereditary basis of variability which is not determined by monomeric differences between genes. For the known recessive or dominant anomalies with regular monomeric inheritance, observations on twins are usually without particular interest. Monozygotic twins behave in this respect simply like a double edition of the same individual, while dizygotic twins behave like ordinary siblings. Observations on twins rarely provide information under these circumstances which we do not already have from family investigations. However, for characteristics which are determined by many genes, family investigations rarely provide a clear picture, and supplementing them by the study of twins is indispensable. On the average, dizygotic twins, like other siblings, agree in half of their genes, while monozygotic twins have all genes in common. If it is found, therefore, that monozygotic twins are completely alike in a trait whose normal variability cannot be sufficiently analyzed by means of family investigations, while dizygotic twins show marked differences, one may conclude that the trait is genetically determined to a great extent. Twin studies, because of their outstanding significance, will be discussed in a separate section.

The peculiarities of polygenic inheritance should not let us forget that the basic laws of heredity are always the same, no matter whether one is dealing with dominant or with recessive genes having a recognizable single effect, or with "polygenic" inheritance. A polygenic system can be understood most easily if we think of it as being constructed of the same constituents that we have already come to know as individual genes. Polygenic inheritance is not a special type of inheritance but a special case of gene action. Let us assume a series of allele pairs all of which influence the same phenotypic property, in each of which one allele has an influence of strength 1, the other an influence of strength 2, and whose effects are additive. We can then develop schematically the additive effect of these genes for the cases of monomeric, dimeric, trimeric, up to polymeric inheritance. If two alleles are involved and if these have the same frequency, the probability that an individual is homozygous for gene a^1 is 0.25; of course, the probability of homozygosity for a^2 is of the same magnitude, while heterozygosity is to be expected with a probability of

0.50. Now, if an additional pair of alleles, b^1 and b^2, participates in the expression of the trait, also with effects 1 and 2 and with equal frequency, the probability of the double homozygosity $a^1a^1b^1b^1$ is only 0.0625 (0.25 $\times$ 0.25); likewise, the probability for the double homozygosity $a^2a^2b^2b^2 =$ 0.0625. The probability for the combinations in which effect 1 predominates $(a^1a^1b^1b^2; a^1a^2b^1b^1)$ amounts to 0.25, the probability for the predominance of effect 2 $(a^1a^2b^2b^2; a^2a^2b^1b^2)$ is likewise 0.25, and the probability that both effects are in balance $(a^1a^1b^2b^2; a^1a^2b^1b^2$ or $a^2a^2b^1b^1)$ amounts to 0.375. For three pairs of alleles with effective strengths 1 and 2, the frequency of pure

TABLE 25

ADDITIVE EFFECT OF SEVERAL PAIRS OF ALLELES, OF
WHICH ONE ALLELE ALWAYS HAS EFFECT 1, THE
OTHER, EFFECT 2. EQUAL FREQUENCY OF ALLELES IS
ASSUMED

	NUMBER OF EFFECTS		COMBINED EFFECT	FREQUENCY OF COMBINATION
	Effect 1	Effect 2		
1 locus	2	0	2	0.25
	1	1	3	0.50
	0	2	4	0.25
2 loci	4	0	4	0.625
	3	1	5	0.25
	2	2	6	0.375
	1	3	7	0.25
	0	4	8	0.625
3 loci	6	0	6	0.0156
	5	1	7	0.0938
	4	2	8	0.234
	3	3	9	0.312
	2	4	10	0.234
	1	5	11	0.0938
	0	6	12	0.0156

effect 1 is only 0.0156, as is the frequency of pure effect 2. There are now five intermediate classifications, in which the effects can combine in the following way:

5 $\times$ effect 1 + 1 $\times$ effect 2; 4 $\times$ effect 1 + 2 $\times$ effect 2; 3 $\times$ effect 1 + 3 $\times$ effect 2; 2 $\times$ effect 1 + 4 $\times$ effect 2; and 1 $\times$ effect 1 + 5 $\times$ effect 2.

In the schematic representation of Table 25, the frequencies of these classifications are indicated.

With each additional pair of alleles which contributes to the effect, the number of classes of differing intensity of effect increases by 2. The frequency of the individual classes is calculated according to the binomial distribution $(a + b)^n$, where n designates the number of alleles, that is, twice the number of loci. The calculation can be carried out simply with the aid of Pascal's triangle, in which the number of members of each series results from the addition of the two numbers above. With an increasing number of participating loci, the approximation to a normal distribution becomes increasingly better. Our example was so constructed that no direction or frequency of individual effects was favored. Thus, deviations in one or the other direction were dependent solely on chance combination. The example was, however, intentionally simplified considerably. If the individual gene effects are unequal, and if their frequencies are also unequal, a normal distribution can nevertheless result if one condition is adhered to, namely, that on the whole no direction or frequency is favored, that is, that the distribution of the genes themselves again obeys the laws of chance. If numerous genes interact, each of which exerts only a small effect, then even the simultaneous influence of a single gene with a stronger effect—in our imaginary example perhaps with effect 3—will no longer be easily recognizable. Its contribution will not

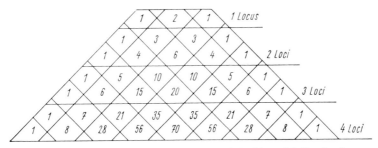

Fig. 63.—Pascal triangle for the derivation of the binomial distribution

appreciably alter the shape of the normal distribution. Thus we have seen that a normal distribution can be built up under very simple assumptions from elementary components of equal intensity and equal frequency. But one may not draw the opposite conclusion from this, that every normal distribution is based on such a uniform system. There is a justifiable conclusion, however: If a feature which has little environmental dependence follows a normal curve in its quantitative distribution, it is to be assumed that it is determined by numerous genes. The scheme used as an example also shows, however, that a rather good approximation to a normal distribution can be obtained with three pairs of alleles.

There is no example of polygenic inheritance in man which actually has been analyzed by a satisfactory method. This is simply due to the fact that the methods used to find the elucidation of polymeric inheritance are not applicable to man. On the positive side, it can be determined that the numerous investigations regarding body size, body shape, variability of physiological data, giftedness, and susceptibility to disease have never given any cause for a departure from the concept of a polymeric basis of normal variability. The first attempt at an analysis of polygenic inheritance in man consisted in the research done by Gertrude and Charles Davenport on the skin color of racial mixtures of black and white persons in the Bermuda Islands and in Jamaica. The skin color of the first generations of hybrids was intermediate between those of the parental races. In the second generation, that is, in the descendants of two hybrids, the spread in color was considerably greater, but there was no simple division with one-fourth whites and one-fourth blacks, as would be expected if the differences in skin pigmentation were determined by a single pair of alleles. Nevertheless, among thirty-two children of the second generation, three could be classified as white. This indicates that not too many loci can be involved in the determination of skin color, for otherwise the probability would be too low that all alleles for white would come together again in the second generation.

The nature of polygenic inheritance is best understood if it is clear that each child has obtained one half of its genes from one parent and the other half from the other parent, and that siblings, on the average, have inherited half their genes in common from their parents. Of course, it need not be true in an individual case that someone has just one-half of the genes which his brother has inherited from the parents. There may be more or fewer of them according to chance. On the average, every grandchild has inherited 25 per cent of his genes from each of his four grandparents. But here, also, chance deviations are possible. Exactly half of the genes are passed on from the grandfather to the father, and again exactly half from the father to the son; but among the genes which are transmitted from the father to the son, although those coming from the grandfather and those from the grandmother on the average are present in equal number, accidental deviations can occur, so that in one instance more hereditary factors from the grandfather, in another instance more hereditary factors from the grandmother, reach the grandchild. If there were no crossing over and if, as a result, the hereditary material were usually passed on in larger packages, deviations from the average condition would be more significant. It could then be expected that physiognomic resemblance between grandfather and grandchild would also be an indication of similar psychological characteristics or similar susceptibility to disease. However, since it appears that crossing over regularly intermingles the genes before every maturation division, significant deviations from chance are hardly to be expected. Therefore, family resemblance in one trait does not tell anything about family resemblance in other traits.

The parent-child relationship and the sibling relationship are the elements of every kinship from which all further conditions of relatedness can be assembled. An uncle is the brother of one parent. As a result, the common possession of inherited genes of parents and children combines with that of siblings: $0.5 \times 0.5 = 0.25$. Uncle and nephew have, on the average, obtained 0.25 of their genes from common direct ancestors. Cousins are children of siblings, and they therefore have $0.5 \times 0.5 \times 0.5 = 0.125$ commonly inherited genes.

The expression "possession in common" of genes must be defined more closely here in order to avoid misunderstandings. Even random "unrelated" persons have a large portion of their genes, the extent of which is unknown, in common. Of course, even "unrelated" persons are related to each other. If one could trace their descent back far enough, one would always come upon common ancestors. On the basis of their relatedness, siblings possess more genes in common than do random persons. How many genes they share depends on the average stock of genes possessed in common by two random persons, and thus on the frequency of the genes in the population. If someone is heterozygous for a very rare gene, the probability that his brother has the same gene is about 0.5. The probability of a random person's having the same gene would be near zero. If, on the contrary, a gene is so frequent that practically everyone in the population possesses it, even siblings may possess this gene in common at most only by a minimal amount more frequently than random persons. Thus, one cannot simply say that siblings have half their genes in common. In actual fact, they have a considerably higher percentage in common.

The concepts of similarity and difference become meaningful only when they are referred to a specific population. In a population of individuals with equal heredity, no greater similarity would be expected between father and son than between two random male individuals of comparable ages. In a population of relatively uniform genotype, only a slightly greater agreement in genotype between father and son would be possible than between random individuals. If, on the other hand, the population is quite inhomogeneous genetically, the significance of the genetic agreement between father and son increases relative to this inhomogeneity. Pronouncements about heredity always have a precise meaning only within the framework of a specific population.

What we mean by resemblance or lack of resemblance is basically the degree of resemblance between persons, measured by the average resemblance between two random individuals. Measured by their lack of resemblance to average Europeans, Japanese or Chinese persons often seem to us to have a startling resemblance to one another. East Asians have very much the same experience with Europeans.

In the case of quantitative traits, the resemblance between two individuals is measured by the correlation coefficient. It measures the differences between individuals to be tested for resemblance on the basis of the differences among arbitrary individuals in the population to which they belong. A certain degree of understanding of the correlation coefficient is indispensable for understanding polygenic inheritance. The concept of correlation will be explained by means of the example of body height of fathers and sons. Figure 64 gives data on the body size of Italian men from two generations. It is a correlation chart in which the position of each father-son pair is determined by the body size of the father on the abscissa and that of the son on the ordinate. Thus, if the number 15 is entered in the square for the paternal size range of 165–69 cm, and for the size range of the sons of 175–79 cm, this means that 15 of the fathers of 165–69 cm size had sons in the size range of 175–79 cm. If the body size of the fathers were the same as that of their sons in each case, numbers would be entered only in the diagonal from the lower left to the upper right. We would then have a positive correlation of +1. If, on the other hand, the size of the sons were completely independent of the size of their fathers, that is, if, for small fathers, the small size classes for the sons were not occupied preferentially and, for large fathers, the large size classes for the sons were not occupied preferentially, no correlation would exist. The correlation coefficient would be 0. Finally, a negative correlation would mean that small fathers more frequently had tall sons, while tall fathers more frequently had small sons. Here, the diagonal from upper left to lower right would be occupied preferentially. A look at the correlation chart informs us that a certain correlation between the body height of fathers and sons exists, since the corners at the lower left and upper right are occupied distinctly more heavily than the other two, and since the numbers as a whole are grouped more densely about the diagonal from the lower

		Body Heights of Fathers (in centimeters)								
		Below 155	155–159	160–164	165–169	170–174	175–179	180–184	185 and higher	
Body Heights of Sons (in centimeters)	185 and higher				3	2	1	1		172,0
	180–184			1	7	6	3	2	1	172,0
	175–179		2	8	15	21	15	1	2	170,7
	170–174		13	26	60	59	20	4	1	168,7
	165–169	4	12	48	92	41	10	1		166,3
	160–164	8	26	37	48	18	9	1		164,1
	155–159	4	16	21	22	8			1	163,2
	Below 155	1	1	4	5	2	1			165,2
	Average Body Height of Sons									Average Body Height of Fathers
		158,8	163,8	165,3	167,4	169,9	171,1	174,7	172,6	

FIG. 64.—Correlation table for body height of Italian men in two generations. (From Boldrini.)

left to the upper right. We can also recognize the positive correlation from the average values of body heights of sons of fathers in the different body-height classifications in the lowest line of the chart, as well as in the last column which indicates the average body height of the fathers of sons in the different body-height classifications. With increasing body height of fathers, the body height of the sons also increases, although not in the same measure. Fathers of 155–59 cm height, that is, with an average height of 157.5 cm, have sons of 163.8 cm height, while fathers of 175–79 cm height have sons of 171.1 cm height. Thus, an increase in body height of 20 cm in fathers is accompanied by an increase of only 7.3 cm in the sons. This "regression" of the average body height of sons compared to their fathers is based on the fact that the body height of the sons is dependent not only on the paternal hereditary material, and the body height of the fathers not only on those genes which they have transmitted to their sons. The fathers also possess the 50 per cent of additional genes which they do not have in common with their sons, and the sons have obtained 50 per cent of their genes from the mother's side. If one starts with the size classifications of the sons and determines the corresponding size of the fathers, as was done in the last row of the correlation chart, one finds, in reverse, a regression of the paternal size relative to that of the sons, which is of about the same order of magnitude.

If a completely regular connection existed between the size of the fathers and of the sons in the sense that one could accurately predict the size of the son from the size of the father, the correlation coefficient would equal +1. If, however, no connection existed, that is, if the size of the sons varied entirely independently of that of the fathers, the correlation coefficient would equal 0. On the other hand, if one could estimate the size of a son from the size of his father with a certain probability, the correlation coefficient would have a positive value between 0 and 1, depending on the "rigidity" of the relationship. A negative correlation coefficient would indicate that small fathers had tall sons and tall fathers had small sons more frequently than is to be expected in a random distribution.

Father and mother transmit an equal number of autosomal genes to their children. If, in polygenic inheritance by additive genes, X-chromosomal genes are involved to a negligible extent, the variability must be influenced by both parents in the same way. Pearson and Lee found the following correlation coefficients for the body height of parents and children:

Mother and son	.49	Mother and daughter	.51
Father and son	.51	Father and daughter	.51

Among adult brothers, correlation coefficients for body height of +.47 (Howells), +.51 (Pearson and Lee), and +.57 (Bowles) were found. The correlation is of about the same magnitude as between parents and children. This is to be expected if the variability of body size is essentially determined by heredity and if the genes which determine body size have an additive effect—but only under the additional assumption that no homogamy exists in regard to body height. Homogamy means that marriages are entered into by persons of similar genotype more frequently than would be expected in a random mixture of the population ("panmixia"). But for body height the proverb applies: "Equal attracts equal." Especially very small and very tall persons find marriage partners more frequently among people of similar body height. For body height, a correlation of $+.20$ to $+.30$ exists between marriage partners. The effect of homogamy on the correlation between parents and children depends on the length of time over which homogamy has been prevalent. With continuous strict homogamy, that is, with a high correlation of the marriage partners in regard to body height, within the population different partial populations would gradually be bred in which marriages would occur and which would be very similar within themselves genetically, but which would be in lesser agreement with the other partial populations. In homogamy, the correlation between parents and children also becomes considerably greater than in panmixia. An extreme example of homogamy with respect to skin color would exist if two races with extreme differences in skin color lived side by side in a country, but if all intermingling were effectively prevented by law and custom. In this case, the correlation between parents and children with regard to skin color would approach the extreme value of $+1$, that is, all children with skin color A would come from parents with skin color A, while all children with skin color B would come from parents with skin color B; the value 1 would not be reached entirely only because of the parent-child correlation within the two groups which differs from 1. If the barriers now fall and a mixture of the population results, the correlation coefficient will gradually drop and finally decrease to $+.50$, assuming that the skin color differences are polygenic and determined by additive genes. The alteration of the correlation coefficient here would have nothing to do with a change in the hereditary influence of the genes, but only with a change in the degree of homogamy in the population. If it is assumed that homogamy with respect to body height exists in a population for only three generations, and that its extent is characterized by a correlation coefficient of $+.25$, one would find, in the case of exclusive genetic determination of the variability in body height by intermediate polygenic inheritance, a correlation of $+.63$ between parents and children in place of the expected value of $+.50$ for panmixia.

EFFECT OF DOMINANCE ON CORRELATION BETWEEN RELATIVES

In the considerations presented thus far on the subject of polygenic inheritance which is responsible for the correlation between parents and children, it was assumed that the polygenes have an additive or intermediate effect. If they are pairs of alleles of which one is always completely dominant and the other recessive, the correlation between parents and children becomes smaller. For genes with intermediate effect, each individual gene participates in the total effect; therefore, the correlation between parents and children or between siblings is determined solely by the common possession of genes. In dominance, the correlation between nearest blood relatives becomes less, since not all genes possessed jointly now play a decisive part; in the heterozygous state and in the homozygous state for the dominant allele, only the one dominant allele is effective, and the recessive allele becomes effective only in the homozygous state. In heterozygotes, the common possession of the recessive alleles is not expressed in the phenotype. The common homozygous occurrence of recessive genes is more frequent in siblings than in parents and children. In the case of polygenic inheritance which is half dominant and half recessive, therefore, a higher correlation is to be expected between siblings than between parents and children. The effect of dominance on the correlation coefficients depends on the frequency of the dominant alleles. If the dominant alleles are rare, the correlations approach those which are to be expected for pure intermediate inheritance; if they are frequent, the correlation decreases, more strongly so between parents and children than between siblings. However, one may not always conclude that dominance exists when there is a higher correlation between siblings than between parents and child, for siblings usually grow up in more similar surroundings than do parents and children. Correlations determined by environment would therefore also be higher for siblings.

TABLE 26

THEORETICAL CORRELATIONS BETWEEN RELATIVES
UNDER DIFFERENT CONDITIONS OF GENE
EFFECT AND OF HOMOGAMY*

RELATIONSHIP	PANMIXIA		HOMOGAMY FOR THREE GENERATIONS (correlation of marriage partners of +.25)
	Heterozygotes Intermediate	Complete Dominance†	Heterozygotes Intermediate
Parent-child...........	.50	.33	.63
Siblings..............	.50	.42	.63

* From Hogben and Stanton, cited according to Tanner.
† For the case of complete dominance, equal frequency of the alleles is assumed.

If the dominant and recessive alleles are of equal frequency, the sibling correlation is depressed to .42 by the dominance effect. If the dominant genes attained a frequency of 90 per cent, the sibling correlation would amount to only .3. In Table 27 some empirically determined correlation coefficients of body measurements between brothers (adults) are cited.

On the basis of these statements, it is incorrect to regard differences in correlation coefficients as proof of stronger or weaker hereditary determination of individual features. Presumably, the low correlations for face width and nose width are based not on a smaller genetic influence on these features, but on the fact that relatively frequent dominant genes are involved in their expression.

The problem of complete dominance of polygenic factors is of some interest in the case of body height. During the past 100 to 150 years, the average body height of young adults has increased in many countries by five to ten centimeters. Some authors have suspected that this remarkable phenomenon could be explained by the so-called hybrid vigor. They say that, with increasing traffic, the former regions of inbreeding, or isolates, have been dissolved. Therefore dominant genes occur in the heterozygous state more frequently than before. The argument can be represented schematically in the following way: Let us assume that, in the parent generation, the genes occur mainly in the homozygous state and that they are different alleles in different populations. Because of the intermingling of the populations, the genes become heterozygous in the children's generation. If the effect of the different heterozygous pairs of alleles is additive, the total effect of the genes must be stronger in the heterozygous state than in the two original homozygous states, since the dominant gene effect is distributed over more individuals while the genes do not lose intensity, owing to their dominance.

TABLE 27

CORRELATION COEFFICIENTS OF BODY MEASURE-
MENTS BETWEEN BROTHERS (ADULTS)*

Measurement	Correlation Coefficient
Body height......................	.47
Length of tibia...................	.58
Head width.......................	.48
Head length......................	.38
Zygomatic bone width.............	.25
Nose width.......................	.25
Facial height.....................	.59
Ear length.......................	.27

* From Howells.

The scheme which presupposes dominant action of the alleles, but also an additive effect of non-allelic pairs of genes, looks as follows:

Homozygous parent generation:

Genotypes:	AA	bb	CC	dd		aa	BB	cc	DD
Gene effect:	2	1	2	1		1	2	1	2

Additive total
effect of genes: 6   6

Children's generation:			Aa	Bb	Cc	Dd
Gene effect:			2	2	2	2

Additive total
effect of genes: 8

The sparse data available on correlations of body height of parents and children and of siblings show no distinct dominance effect. Thus the hypothesis that the increase in height of the population is determined by this hybrid vigor lacks its decisive foundation. Even in hybridization between human races of very different origin, where one could most readily expect differences in homozygous genes in the original races, no general dominance of body height has been shown, but instead an intermediate behavior.

If one calculates the correlations between more distant relatives, one must take into account the decreasing steps of frequencies of common genes. Parents and children as well as siblings have 50 per cent of their genes in common, uncle and nephew, 25 per cent, cousins, 12.5 per cent. In polygenic determination and intermediate gene action, the correlation coefficients are then reduced from .5 to .25 and .125. In still more remote relationships, the common gene property becomes so low that it can hardly be determined by means of correlation coefficients. Wingfield found certain correlations in tested intelligence between different relatives (see Table 28).

Here, the correlation between cousins is still conspicuously high compared to the correlation between siblings or between parents and children. By pure hereditary determination, it would be expected that the correlation between cousins would be only one-fourth as large as that between siblings. But the simple theoretical assumptions for the model are again sensitive to the perturbation introduced by homogamy. In regard to intelligence, a pronounced "elective affinity" between marriage partners exists. This is so significant that marriage partners are usually more alike in their intelligence level than siblings. Outhit has calculated a correlation of intelligence of +.74 between marriage partners. Especially very intelligent persons and those of very low intelligence usually marry partners with whom they are at about the same

TABLE 28

CORRELATION IN TEST INTELLIGENCE
BETWEEN RELATIVES

Relationship	Correlation Coefficient
Identical twins	.90
Siblings	.50
Parents and children	.30
Cousins	.27

level in regard to intelligence. Therefore, the level of intelligence is approximately equal even in rather wide circles of relatives, although this cannot be explained by jointly inherited genes. The absence of a distinct decrease of correlation coefficients with decreasing degree of relatedness can thus be an indication of homogamy.

The absence of a lowering of correlation coefficients with increasing remoteness of the relationship, however, can also be due to the fact that the social and economic environment is usually similar within wide circles of relatives. "Homogamy" also exists with respect to property, education, and social status. Therefore, characteristics depending on the social milieu also show relatively high correlations between more distant relatives. The correlation coefficient is by no means a pure measure of hereditary influence. Environmental factors common to a family can also cause traits with environmental dependence to show a family resemblance. Burt and his collaborators were able to study the development of a large number of children who were placed in orphanages and children's homes during their first weeks of life and who had grown up under rather uniform conditions. These children, who had been raised together but were not related to one another, showed correlations of +.25 and +.27 in their achievement in two intelligence tests. One might think that this measure of agreement is determined only by the common environment. But this, too, is questionable. These are, for the most part, children of unmarried mothers; extramarital motherhood occurs more often among unintelligent than among intelligent girls. A part of the agreement in the test achievement could be due to the fact that these children did not constitute a representative random sample of the population which was distinguished only by its common environment, but that they also agreed somewhat more than random children in the genetic basis of their intelligence. One of the most impressive results of the investigations by Burt, and of a similar study by Burks of 214 foster children, was the occasional occurrence of high intellectual ability among the children in the homes. In most cases it was found that the fathers of these children also had above-average intelligence.

We may summarize up to this point: Correlation calculations alone do not permit a quantitative determination of hereditary and environmental influences. Even with complete environmental independence of a trait, the correlation calculation does not provide any clear-cut or even generally valid results. It involves several unknowns, the most important of which are the degree of dominance, the frequency of the dominant genes, and the extent of homogamy. The correlation calculation is thus in need of refinement, which would make possible a better separation of hereditary and environmental influences. We shall learn more about this from twin studies.

Twin studies

The great scientific value of twin studies is due to the fact that there are two different kinds of twins: monozygotic and dizygotic twins. Monozygotic twins come from a single fertilized ovum and result from its subsequent division. This process is comparable to the asexual propagation of some species of plants and animals. Two genetically alike individuals are created from one individual. In sexual propagation, on the other hand, every newly conceived individual is also an individual genetically.

Dizygotic twins are distinguished from ordinary siblings only by their common time of conception and thus by the common pregnancy. In their hereditary traits they are not more similar than ordinary siblings. Like ordinary siblings, they come from two different maternal ova fertilized by two different sperm cells.

Twin studies compare monozygotic with dizygotic pairs of twins. In both types of twins, the partners usually grow up under the same, or at least under very similar, conditions. Therefore differences which appear in dizygotic but not in monozygotic twins can be ascribed to differences in their hereditary traits. Conversely, differences which are present in both monozygotic and dizygotic twins must be traced back to environmental conditions. To be sure, these environmental conditions are usually difficult to determine. They may involve small accidental fluctuations in the intra-uterine environmental conditions which can have far-reaching consequences. Thus, it can happen that one of a monozygotic pair of twins has a cleft lip and palate while the other is entirely normal. The usual type of twin study thus investigates the differences between twins. Whatever is different in dizygotic twins in spite of their common environment, but is the same in monozygotic twins, can be designated as hereditary.

Another important question posed in twin studies starts with monozygotic twins in different milieus rather than with twins in the same environment. Here one observes which features remain the same in spite of the different environment and which develop differentially. In this case, the twin method is really not a genetic method in the strict sense, but a special form of environmental research. The influence of exogenous factors is observed. This can also be done with subjects who are not twins. But in the case of twins, one has at hand the uniquely favorable prerequisite that every observed case is accompanied by a genotypically identical control case. Observations on monozygotic twins who have grown up separately are of special interest, as they provide information about the extent to which human characteristics are modified by differences in environment, or the extent to which they remain the same in spite of the different surroundings. Ordinary twin studies of pairs that have grown up together do not touch upon this question.

Finally, monozygotic twins have also been exposed intentionally to differing conditions, for instance by attempting to advance only one of them in infancy by means of calisthenics and crawling exercises and considerable attention, while leaving the other mainly to himself. Through this so-called co-twin control it was possible to demonstrate that the psychomotor development in infancy is primarily a maturation process, which occurs largely without dependence on external stimuli. A similar principle of co-twin control is the basis of observations of pairs of twins with brain trauma. By means of electro-encephalographic, neurological, and psychological examinations and by means of a follow-up of the later fate of persons with brain injuries who had uninjured monozygotic twins, it was possible for the first time to obtain a reliable and precise evaluation of the relatively small effect even of severe brain trauma. Twin control also offers some favorable means for the testing of therapeutic procedures. Some few observations of the result of treatment in only one partner of a pair of monozygotic twins could supply more valuable knowledge than can be obtained from a large quantity of other statistical information, in view of the genetic multiplicity of the ordinary population and thus the limited possibility of comparing one case with another.

Frequency.—On the average, there is one twin birth for every 85 to 90 births. Thus, 1.1–1.2 per cent of all births are twin births and, correspondingly, 2.2–2.3 per cent of all newborn infants are twins. Since twins usually are born prematurely and with deficient weight, they have a higher infant mortality than do other children. In later childhood, therefore, only about 0.9 pairs of twins both of whom are living are found per 100 children.

Among all pairs of twins, about a fourth to a third are monozygotic, as can be calculated approximately from the sex ratio of twin births. Monozygotic twins are always of the same sex, since they have originated from a single fertilized egg and, as a result, agree in their sex chromosomes just as in the remaining chromosomes. Dizygotic twins can be of the same or of opposite sex; the probability is the same that they will be of identical or of opposite sex, if one neglects the fact that slightly more boys than girls are born. Among dizygotic twins, then, equal numbers of the same as of the opposite sex would be expected. In Hamburg, a total of 1,713 pairs of twins were born between 1946 and 1954; of these, 588 were of different sex and therefore certainly dizygotic twins. In addition, an equal number of dizygotic twins of the same sex is to be expected. The total number of twins of the same sex amounted to 1,125; if we subtract 588 from this, we obtain 537, the number of monozygotic pairs. Among all pairs of twins, 31.3 per cent would be monozygotic. The percentage of monozygotic twins among all twins is not a biological constant. It decreases with the age of the mother, since births of dizygotic twins become more frequent with increasing age of the mother up to a maximum at between 35 and 40 years, but then again become less frequent. In northern countries, the frequency of dizygotic-twin births is greater than in southern countries. The frequency of births of monozygotic twins, on the other hand, shows no pronounced geographical differences. A knowledge of the normal ratio of monozygotic to dizygotic twin births is important for the evaluation of twin studies. Monozygotic twins are more conspicuous and are therefore included preferentially in every collection of twins which is not strictly free of selection. However, since not the monozygosity as such, but rather the degree of resemblance is the decisive factor here, one must assume that material which contains a disproportionate number of monozygotic twins is selected on the basis of resemblance. In such material, presumably, the dizygotic twins are especially similar, so that it is unsuited for conclusions which demand more general validity, especially for quantitative statements.

Diagnosis of monozygotic and dizygotic twins.—The obstetrical diagnosis of "monozygotic" or "dizygotic" twins concerns the condition of the membranes, primarily the chorion. Twins with separate chorions are designated as dichoric and twins with a common chorion, as monochoric. These findings are not in agreement with the genetic distinction between dizygotic and monozygotic twins. It is true that twins with a common chorion are always monozygotic also in the genetic sense, apart from very rare exceptions which evidently are caused by secondary disorders of the membranes; however, not only all dizygotic, but also the majority of monozygotic twins have separate chorions. Whether monozygotic twins have a common chorion or two chorions appears to depend merely on the time of division of the originally single ovum. If the division occurs before implantation, the monozygotic twins will have separate placentas and thus, of course, also separate chorions. It would be of great assistance for twin studies if the condition of the membranes were carefully recorded in obstetrical hospitals.

The genetic diagnosis of monozygotic and dizygotic twins (similarity diagnosis) is an exclusion diagnosis. If a pair of twins shows differences which are known to be genetically determined, they are not monozygotic. First, all twins of unlike sex are dizygotic; also, all pairs which show marked differences in any one of the numerous physical features which are known to be determined by heredity. Features which are considered reliable criteria are structure and color of the iris, form and color of hair, freckles, ear shape, hairline, hairiness of fingers, shape of the nose and mouth, shape and size of the teeth, shape of fingernails, shape of hands, dermal ridges of the fingertips and of the palm, body size and proportions. The certainty of the decision increases with the number of features taken into account. It is impractical to adhere to too rigid a scheme, since it is often the rare, conspicuous features which are of special value. In the majority of all cases, a distinction on the basis of the physical features is possible without difficulty. Sometimes, however, even dizygotic twins resemble each other so much that they can be mistaken for monozygotes. Monozygotic twins are less often mistaken for dizygotes; this seems to have happened especially when only one of them had a malformation or disease, of whose hereditary cause the author in question was convinced.

In all cases which are in any way doubtful, it is desirable to make use of the antigens of the blood corpuscles as objective features whose hereditary determination has been proved. The blood group polymorphism of the population is so great that dizygotic twins only rarely show agreement in all blood antigens. For clearly distinguishable conditions in the polysymptomatic morphological comparison, and thus for the majority of all pairs of twins, blood factor determinations would constitute a waste of time and money. However, they are important in individual cases which are of special theoretical significance. These have included, for example, observations of concordant and of discordant Ullrich-Turner syndrome and of Klinefelter's syndrome in a pair of monozygotic twins each. Skin transplants between monozygotic twins can also serve as a very sensitive criterion for the agreement of numerous hereditary factors. In the pair of twins with Klinefelter's syndrome, the transplants were accepted without reaction. In a famous case where two pairs of twins were born in a hospital on the same day and one partner of one pair was accidentally exchanged with one partner of the other pair, Franceschetti, as legal-medical expert, had a skin graft carried out. The complete acceptance of the transplants between the two brothers who had grown up in separation and who had reached the age of six years, and the rejection of the transplants between the two boys who had grown up together but evidently came from different parents, was one of the most impressive proofs of the genetic identity of the separated pair. However, a positive diagnosis of monozygosity cannot be completely assured even on the basis of skin grafts. It could also be possible that, at some time, a pair of dizygotic twins agree in all genes which are responsible for histo-incompatibility. In 1927, the surgeon K. H. Bauer made the possibility of transplants between monozygotic twins therapeutically useful for the first time. He was successful, with a monozygotic pair of twins having a concordant syndactyly of the ring finger and little finger, in using a skin flap from one partner for covering the surgical wound in the other. Even transplants of whole kidneys and of thyroid tissue have been carried out successfully with monozygotic twins.

Twin studies are valuable primarily for answering the question whether hereditary factors are at all involved in the individual differences of normal or pathological features. If monozygotic twins do not agree considerably more often than dizygotics, hereditary factors obviously do not play a large role. Thus, it was proved by means of twin studies that the hereditary trait has only little significance in cases of multiple sclerosis or of cancer.

Thums found among 13 monozygotic pairs of twins, one of whom had multiple sclerosis, only one case in which the twin brother also had multiple sclerosis. Of 30 dizygotic pairs, one had a questionable concordance. A paper by Curtius shows how this picture can change if one compiles case histories from the literature instead of using a series which is free of selection. In this way, he found 13 concordant to 12 discordant monozygotic pairs of twins. It is true that, for most of the concordant pairs, the data are insufficient for a definite clinical diagnosis. It is to be noted that, in the collected statistics of Curtius, also, all pairs of twins who constituted parts of a larger published series of cases were discordant. Recently MacKay and Myrianthopoulos were able to collect data on 54 pairs of twins with the aid of the National Multiple Sclerosis Society. Of 29 monozygotic pairs, 2 were definitely found to be concordant when the strictest diagnostic criteria were applied. Of 25 dizygotic pairs, 1 pair was concordant. When neurological anomalies were included which were compatible with, but not sufficient for, the diagnosis of multiple sclerosis, 7 of the monozygotic pairs and 4 of the dizygotic pairs were found to be concordant. These figures can hardly be reconciled with a pronounced hereditary influence.

Twin studies and cancer.—There are several rare hereditary anomalies which frequently lead to malignant degeneration. In diffuse polyposis of the colon, which has dominant inheritance, carcinoma of the colon and rectum almost always results. A special form of the dominant hyperkeratosis palmarum et plantarum accompanies carcinoma of the esophagus. In the long run, the skin changes of recessive xeroderma pigmentosum always deteriorate in individual locations into malignant ones. On the whole, however, heredity plays a small role in cancer. Malignant tumors are so frequent that, even in random distributions, families with numerous cases of cancer death are not infrequently observed. Family studies are complicated by the fact that, at the relatively advanced age at which the different types of cancer usually occur, it is seldom still possible to obtain reliable data about the preceding generations. Twin studies have supplied decisive contributions in this field. Verschuer has given a summary of the literature and of his own extensive series in Table 29 (Verschuer and Kober, 1956).

If we disregard the location of the tumor, the agreement between monozygotic pairs of twins (19 per cent) is not much higher than between dizygotic pairs of twins (15 per cent). However, the monozygotic twins show much closer agreement in the type and location of the tumor. Verschuer concluded from this that a hereditary influence is clearly shown in regard to tumor location, but not in regard to the illness itself.

TABLE 29

SERIES OF TWINS WITH CANCER

	n	kk	kd	d
Monozygotic twins........	129	20	5	104
Dizygotic twins..........	287	7	37	243

n = number of pairs of twins.
d = discordant: the proband has cancer, the partner has not.
kk = concordant in kind and location of tumor.
kd = both partners have cancer but of different kind and location.

If the type and location of malignant tumors show a hereditary influence, it is difficult to understand why this hereditary influence does not appear in the total number. We must therefore examine the figures more closely in order to understand this contradiction. It is particularly strange that tumors differing in location occur so much more frequently in dizygotic twins than in monozygotic pairs, namely, in 15.2 per cent as compared 3.9 per cent. If hereditary factors were responsible for this, they would have to cause a higher total concordance in monozygotic twins. If, however, hereditary factors played no role, it is not apparent why discordant locations should not occur as frequently in monozygotic as in dizygotic twins. Apparently the higher figure for the tumors with concordant occurrence and discordant location in dizygotic twins can be explained by two factors. First, the dizygotic twins also include pairs of unlike sex. Almost half the cases involve breast or uterine cancer. In twin brothers of female patients with breast or uterine cancer, concordance in regard to location is impossible. Of course, no special hereditary trait for a special cancer location is responsible for this, but simply the difference in sex, which is determined by heredity. To this extent, hereditary factors are responsible for tumor location. The figures in Table 29 could be better evaluated if twins of unlike sex had been omitted. A part of the concordance in regard to occurrence and location of tumors may be determined by chance, because of the frequency of malignant tumors. However, this chance-determined part can be compared in monozygotic and dizygotic twins only if the pairs of twins are of like sex. Twins of unlike sex constitute about half of all dizygotic pairs of twins. The effect of the sex difference thus carries considerable weight. Still, probably not the entire difference between the 15.2 per cent and the 3.9 per cent for the discordant location can be explained by this. The remaining difference could be due to the fact that the dizygotic pairs of twins, on the average, became ill 4.7 years later than monozygotes. At the average age of onset of illness of the twins, which was

53.5 years for the dizygotic and 48.8 years for the monozygotic pairs, an age difference of five years means an increase in the frequency of cancer by 40–50 per cent.
Thus the greater frequency of tumors with discordant locations among dizygotic
twins is probably due in part also to the fact that they were counted at an age
at which the danger of cancer is considerably greater. If there were a sufficient basis
for making the cancer frequency in monozygotic and dizygotic twins comparable
by standardizing the age, the total concordance rate for the monozygotic twins
would increase. Thus, in fact, the two series are not comparable in regard to the
important factors of sex and age. If one takes this into account, one must reach
somewhat divergent conclusions. However, the observation remains valid, regardless of the above mentioned uncertainties, that hereditary factors play a subordi-
nate role in cancer.

Twin studies and tuberculosis.—The clinical course of tuberculosis shows
great individual differences. These depend partly on sex and age, partly on
diet and other living conditions, partly on the type of infection, but certainly also on the genotype of the patient. Only the twin studies of Diehl and
Verschuer have made possible a clear judgment regarding genotypic differ-

TABLE 30

PREDISPOSITION TO TUBERCULOSIS IN TWINS

	n	Concord-ant	Per Cent
Monozygotic twins........	381	202	53.0
Dizygotic twins..........	843	187	22.2

ences in the predisposition to tuberculosis. Meanwhile, numerous other
series of twins have substantiated these basic results. Table 30 gives a summary of the observations up to date.

In this material, differences in the predisposition of twins to tuberculosis
meant considerable differences in the clinical course when both twins were
infected, as well as the infection of one partner only when the other remained well. Thus the 53 per cent concordance for monozygotic twins is a
minimum figure which does not reflect the concordant behavior with respect
to the actual occurrence of tuberculosis, but only the concordance rate of all
monozygotic pairs of twins at least one of whom had tuberculosis, without
consideration whether the other was infected at all. If only those monozygotic twin pairs had been included in which both partners were infected, the
concordance rate would, of course, have had to be higher. Of course, complete freedom from the disease probably is also determined in part by
hereditary factors. At any rate, the percentage of concordant infections is
greater for monozygotic than for dizygotic twins. Table 31 is even more
impressive than Table 30. Here, death due to tuberculosis or cure among
pairs of twins who were concordant in the disease was chosen as the criterion
for concordance or discordance, rather than freedom from or affliction with
the disease and the clinical differences in its course. These are the results of

examinations in the series by Diehl and Verschuer, carried out twenty years later.

Thus, for pairs of twins both of whom were originally ill, after twenty years concordance in regard to the outcome in either death or cure was found in 13 of 19 monozygotic pairs of twins, and in 4 of 20 dizygotic pairs. However, in a larger series of 73 pairs of twins with concordant tuberculosis, Kallmann and Jarvik also observed 9 dizygotic pairs both partners of which had died, and 13 monozygotic pairs of twins of which only one partner had died.

The considerations which we came to know in the evaluation of the frequency of a disease among the siblings of probands are equally valid in the evaluation of the concordance frequency of a disease in twins. In the special case of twins, this is not always thought of, although it is immediately plausible. Let us assume that we detect a rare hereditary disease with regular

TABLE 31

FOLLOW-UP EXAMINATIONS OF TWINS WITH TUBERCU-
LOSIS OF BOTH PARTNERS*

	Both Deceased	Both Clinically Healthy	One Deceased, One Clinically Healthy
Monozygotic twins.........	5	8	6
Dizygotic twins...........	...	4	16

* From Mitschrich.

dominance in dizygotic twins. The probability that a random child, only one of whose parents is heterozygous for the hereditary disease, has the hereditary disease amounts to 50 per cent. The probability that a dizygotic pair of twins is concordantly afflicted amounts to 25 per cent, the probability that both are well, also 25 per cent, and the probability that only one is afflicted, 50 per cent. Thus, of all dizygotic pairs of twins who can be detected because at least one is afflicted, only a third are concordant. But this is true only if all pairs of twins in the population are included, or at least if concordant pairs of twins are not included with a higher probability than discordant ones. If the detection of a concordant pair of twins is twice as probable as that of a discordant pair, we obtain directly the theoretical concordance frequency of 50 per cent. If discordant pairs are included as frequently as concordant ones, a correction must be made according to the sibling method, that is, the concordant pairs must be counted starting with both partners of each pair. In this way, we also obtain 50 per cent. In twin studies, too, the method of collection of material must be noted and taken into account in the evaluation. If twins are collected more or less at random as one happens to observe them, one cannot expect to be able to draw very reliable conclusions from the data.

Twin studies and normal features with polygenic inheritance.—Monozygotic twins have all genes in common, and dizygotic twins 50 per cent of their genes, corresponding to the common gene property of ordinary siblings. For exclusively hereditary determination of the normal variability of a feature, one would expect a correlation coefficient of $+1$ for monozygotic twins; but for dizygotic twins, depending on the predominance of intermediate or dominant polygenic inheritance and depending on the degree of homogamy in the population, a correlation coefficient between .3 and .6 would be expected. Thus, the relationship between the correlation coefficients of monozygotic and dizygotic twins is not a measure of heredity in the sense of independence from environment. Table 32 contains data on monozygotic twins, some of whom were raised separately, compared to dizygotic twins and to siblings.

TABLE 32

CORRELATION OF TWINS AND SIBLINGS IN BODY MEASUREMENTS

MEASUREMENT	MONOZYGOTIC TWINS		DIZYGOTIC TWINS	SIBLINGS
	Raised Together	Raised Separately		
Body height.........	.93	.97	.64	.60
Weight.............	.92	.89	.63	.58
Head width.........	.91	.88	.65	...
Head length........	.91	.92	.69	...
Number of pairs.....	50	19	50	26

It is especially noteworthy that monozygotic twins who have been raised separately are considerably more similar in their body dimensions than dizygotic twins who were raised together, and that they show as much agreement as monozygotic twins who grew up together. On the basis of these figures, there can be no doubt that differences between dizygotic twins are essentially determined by heredity and that environmental differences due to the separation of the monozygotic twins in early childhood have no noticeable influence on their physical development. However, one should not draw the more general conclusion from this that differences in milieu are without any significance for physical development. Even for the separately raised pairs of twins the environment was usually quite similar in regard to the social position of the foster parents, schooling, and geographical aspects. Thus the observations only provide information supporting the fact that small environmental differences do not carry any weight compared to differences determined by heredity. Francis Galton, the founder of twin studies, has stated with all necessary clarity that the conclusions about the relationship of hereditary traits and environment which can be drawn from observations on twins are valid only under certain environmental conditions: "There is no escape from the conclusion that nature prevails enormously over nurture when the differences of nurture do not exceed what is commonly to be found among persons of the same rank of society and in the same country." On the other hand, all conclusions from twin studies are valid only for differences between dizygotic twins. These differences are primarily determined by heredity, but they are smaller than the heredity-determined differences among random individuals in the population. In particular, as we have seen, homogamy leads to an increased resemblance among siblings which goes beyond that which one would expect on the basis of the genes inherited in common. In Table 33, the results of intelligence tests for monozygotic and dizygotic twins are shown.

Comparing the first and the third column, we learn that, under equal environmental conditions, the test intelligence depends to a large extent on the hereditary predisposition. In contrast, comparison of the first with the second shows that, for different environmental conditions, distinct differences in the

TABLE 33

CORRELATION OF INTELLIGENCE QUOTIENTS IN TWINS

Test	50 Monozygotic Pairs of Twins, Raised Together	19 Monozygotic Pairs of Twins, Raised Separately	50 Dizygotic Pairs of Twins
Binet...............	.91	.67	.64
Otis................	.92	.73	.62
Stanford educational..	.96	.51	.88

test performance can occur in spite of equal heredity, and that these differences are almost as great as the heredity-determined differences between dizygotic twins.

Attempts at a quantitative analysis of the relationship between hereditary and environmental factors.—If a feature with quantitative variability depends on two different influences at the same time, the total variance of the feature results from the addition of the two individual variances which are determined by the two factors.[2] Variance is the square of the standard deviation. For the standard deviation, which equals the square root of the variance, the additive behavior, of course, is not valid. In twin studies, one can regard the variance within the monozygotic pairs of twins (σ_E^2) as the variance determined by environment, the error in the measurement being at first disregarded. The variance within the dizygotic pairs of twins (σ_Z^2) is composed of the variance determined by environment and that determined by heredity. If we now subtract σ_E^2 from σ_Z^2, we obtain the variance due to heredity. We want to find the contribution of the heredity-determined variance to the total

[2] This is valid only under the assumption that the effects of the two influences occur independently of each other. If an influence a has a different effect under the simultaneous influence b_1 than under b_2, the variance analysis is not applicable.

variance within the dizygotic pairs of twins. This contribution is designated as h^2 or "heritability" and becomes

$$h^2 = \frac{\sigma_Z^2 - \sigma_E^2}{\sigma_Z^2}.$$

In Table 34 some values for heritability are compiled.

Heritability in this special sense is not a measure of the extent to which a feature is determined by hereditary factors but of the portion of the variability of the feature which is genetically determined. This depends on the one hand on the genetic heterogeneity or homogeneity of the population and, on the other hand, on the similarity or difference of the environment. If all individuals in a population agree in the genes responsible for a feature, h^2 becomes 0, even if the feature is determined almost exclusively

TABLE 34

VARIANCE ANALYSIS OF BODY MEASURE-
MENTS OF MONOZYGOTIC AND
DIZYGOTIC TWINS*

Measurement	h^2
Weight.....................	0.69
Body height................	0.88
Head length...............	0.54
Head width................	0.72
Zygomatic arch width.........	0.60
Nose width................	0.66
Facial height...............	0.74

* From Clark.

by these genes. The "heritability" is always valid only for the population in which it was determined. The values of h^2 are not biological constants.

F. Lenz has pointed out a weak point in all quantitative comparisons between dizygotic and monozygotic pairs of twins. This is the error of the measurement. For genetically identical twins who also have not become different because of environmental differences, the error in measurement can only lead to a difference. For dizygotic twins, who are different to begin with, the error in measurement can increase as well as decrease a difference. On the average, therefore, the error in measurement has a different meaning for monozygotic than for dizygotic pairs of twins. This must be kept in mind especially in psychological examinations of twins, in which the error of measurement is relatively large.

It has already been pointed out briefly that the analysis of variance presupposes that the effects of different influences operating simultaneously are independent of one another. If this assumption does not hold, the application of variance analysis is not justified. Unfortunately, it appears that the assumptions of variance analysis are not always fulfilled in twin studies. At least, the possibility that the effect of a certain environmental influence depends on the genotype must usually be considered. For example, one can hardly doubt that the effect of excessive food intake on body weight differs depending on the genotype. In psychology, also, an interaction of heredity and environment rather than an independent effect exists. For persons below a certain intelligence level, attending the best school remains ineffective; it has its full effect only with highly gifted children.

Thus, the only conclusion which remains is that *even upon application of the most perfected statistical procedures and upon consideration of all known contributing factors, all attempts at a quantitative solution of the problem of heredity and environment are doomed to failure.* A number of older attempts at making a quantitative determination of the contributions of heredity and environment without the use of variance analysis are faulty even in their statistical form and therefore will not be discussed here.

Participation of hereditary factors in
common diseases

Most monomeric recessive or dominant hereditary diseases are definitely rare. Among recessive hereditary diseases, sickle cell anemia and thalassemia occupy an exceptional position which presumably is caused by the fact that the responsible gene in the heterozygous state confers an advantage in regard to survival. The next most frequent recessive disease, cystic fibrosis of the pancreas, has a frequency of about 0.1 per cent. The most frequent hereditary ailments with simple dominance occur with comparable frequency, namely, polycystic kidney disease of adults with about 0.1 per cent, and neurofibromatosis with about 0.05 per cent. The frequency of simple monomeric hereditary diseases is determined by the mutation rate and by the effective fertility. The rarity of mutations and decrease in viability due to the disease have the result that simple monomeric hereditary diseases do not have a large share in the total morbidity of the population. In addition to these diseases with established genetic etiology, there are different anomalies which are considerably more frequent; in these, hereditary factors play a certain role, but no satisfactory genetic hypothesis has been found thus far in spite of extensive investigations. Table 35 contains some estimates of the frequency of such diseases.

For *essential hypertension,* Weitz assumed monomeric autosomal dominant inheritance but also pointed out the great significance of diet for the manifestation of the hereditary disease. Among 92 siblings of patients with hypertension, Weitz found higher average blood pressure values in each age group than in control cases. Because of the dependence of blood pressure on age, Weitz evaluated mainly the blood pressure values of the siblings who were older than the patients. Among 47 older siblings who were examined, 18 had blood pressures above 160 mm Hg. Eleven additional siblings had probably died of hypertension. If these were included with the hypertensives among the siblings, there were among the siblings 29 healthy individuals for 29 afflicted ones, so that the ratio for simple dominance seemed to be present. This calculation really applies only to a rare dominant gene. With a hereditary illness of the frequency of essential hypertension, it is very improbable that only one of the parents is hypertensive in each case; instead, one must assume that both parents are hypertensive in a significant percentage of cases. Then the expectation of hypertension among the children would, of course, be higher than 1:1. Pickering has pointed out that the blood pressure values in the total population follow an approximately normal distribution if a logarithmic scale is chosen as measure of the blood pressure. At the age of 60 years, the average systolic blood pressure is approximately 160 mm Hg, that is, it is at the level which is ordinarily chosen as the dividing line between normal blood pressure and hypertension. Between 70 and 80 years of

TABLE 35

FREQUENCY OF SEVERAL DISEASES IN WHICH HEREDITY
PLAYS AN IMPORTANT BUT NOT PRECISELY
DEFINABLE ROLE

Disease	Frequency (per cent)	Author
Essential hypertension.........	30–40	Sobye
Ulcus duodeni et ventriculi.....	6–10	Doll and Jones
Diabetes mellitus.............	2–4	Spiegelmann and Marks
Asthma and hay fever.........	4–9	Ratner and Silberman
Pronounced allergies...........	10	Ratner and Silberman
Mild and severe allergies.......	60	Ratner and Silberman
Schizophrenia...............	1	Sjögren

age, the majority of all people has hypertension in this sense. In no age group, however, is there a double-peaked blood pressure curve such as one would expect if the population could be separated into two sharply distinct groups, those having the gene for essential hypertension and those without this gene.

Pickering therefore developed a hypothesis according to which essential hypertension is not a disease syndrome that can be distinguished qualitatively from the normal state, but rather is only the extreme portion of the normal distribution curve. In this respect, then, it would be comparable to the position of conspicuous tallness within the normal distribution curve of body heights. Pickering and his collaborators were actually able to advance some important arguments in favor of this hypothesis. First, he was able to show that the immediate blood relatives of hypertensives in every age group from the second to the eighth decade of life have blood pressures which are higher by about 15 mm than those of normal persons. Investigations by Miall and Oldham also showed that a correlation of arterial blood pressures between probands and their first-degree relatives exists, which can be designated by a regression coefficient of $+.24$. This regression coefficient is independent of the blood pressure level of the probands. Evidently the same laws for inheritance of blood pressure are valid over the entire scale of possible blood pressure values. To this extent, the observations actually appear to agree best with the explanation of polygenic hereditary determination of the quantitatively variable feature of blood pressure. Pickering wanted to introduce an additional and, if it were sound, doubtlessly decisive argument for this. If hypertension depended on a single dominant gene or even on a recessive pair of alleles, one would expect to find two distinguishable groups of individuals among the siblings of hypertensives, namely, first the carriers of the gene, whose blood pressure values would be scattered about a pathologically increased mean value and, second, siblings who did not possess the gene, whose blood pressure values would be scattered about the normal average. The total distribution curve of blood pressure values of the siblings would then have two peaks. Pickering did not find such a double peak, but instead a uniform distribution curve for the blood pressures of siblings of probands. This is the weakest point in his method of proof.

Weitz had previously found blood pressure values in the siblings of his hypertension patients which were distributed roughly around two peaks in the curve, one in the region of normal average values and one in the region of hypertension. By compiling the blood pressure values of 252 siblings, from 45 to 60 years of age, of patients examined by Sobye and by Oldham, Pickering, Fraser Roberts, and Sowry, Platt was again able to obtain a distinctly double-peaked curve with one peak at 130 to 135 mm Hg and a second at 160 to 165 mm Hg. When he chose as a dividing line between normal and hypertensive individuals a systolic blood pressure of 150 mm Hg, 59 per cent of the siblings turned out to be hypertensives. This value would approximately be expected under the assumption that hypertension has simple dominance and that 20 per cent of the population are hypertensives. Here, 0.80 would be the frequency of individuals who are homozygous for the normal recessive allele (p^2). From this, the gene frequency of the recessive allele would be calculated as 0.894 ($p = \sqrt{0.80}$). The dominant gene would have a frequency of 0.106 ($q = 1 - p$). The heterozygote frequency becomes 0.189 ($2pq$), and the frequency of individuals who are homozygous for the dominant gene, 0.011.

A decision in favor of one or the other hypothesis of heredity is not yet possible at the moment. It would be conceivable that a dominant chief gene is responsible for a considerable part of the variability in blood pressure, but that, besides, additive polygenes are involved which tend to blur the clear picture. The problem is further complicated by the fact that blood pressure is not determined exclusively by genes but also by environment. Even among monozygotic twins of whom at least one had essential hypertension, the second partner was also afflicted only in 50 per cent of the cases. Several significant life histories of monozygotic twins with discordant hypertension point especially to the importance of overeating in the causation of hypertension. Among dizygotic twins, concordance was found in only 23 per cent. If the effect of a dominant hereditary trait were the determining factor, a concord-ance of over 50 per cent would have been expected for dizygotic twins, and of 100 per cent for monozygotic twins.

For *gout,* Hauge and Harvald have discussed polymeric inheritance. They found higher uric acid values in the blood of siblings of gout patients than in control cases, but it was impossible to separate the distribution of uric acid values in the siblings of patients into two distributions, one normal and one raised. For the siblings they found instead a normal distribution dis-placed toward higher values; this was reconcilable with polymeric inherit-ance, but not with monomeric dominant or recessive inheritance. Hauge and Harvald therefore assume that the serum uric acid level is affected in each person by a number of different genes, as are normal characteristics generally.

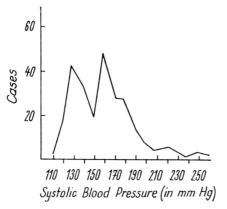

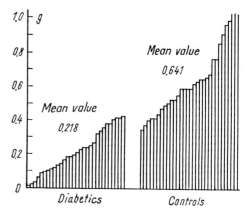

Fig. 65.—Systolic blood pressure of 252 sib-lings of patients with essential hypertension. (Material by Sobye and Pickering. From Platt.)

Fig. 66.—Distribution of weights of β cells. (From McLean and Ogilvie.)

Hereditary hyperuricemia thus can be regarded as an extreme variation of a normal feature. The distinct sex dependence of gout fits easily into this concept. Among 702 gout patients in the literature, only 58 (8.3 per cent) were women. In men, the uric acid level in the blood is about 1 mg per cent higher than in women. Therefore considerably fewer women reach the threshold value which leads to gout. Evidently, endocrine influences are re-sponsible for this difference between the sexes, since even men with hered-itary hyperuricemia develop pathologically raised uric acid values only after puberty.

The results of Smyth, Cotterman, and Freyberg for 87 immediate relatives of gout patients deviate somewhat from the results of Hauge and Harvald. They indicate instead a two-peaked distribution of uric acid values among blood relatives and thus a monomeric dominant hereditary determination of hyperuricemia. Among male relatives before puberty, and among some of the female relatives, the effect of the gene was not recognizable, but in adult men it appeared regularly to lead to hyperuricemia.

In *ulcus duodeni,* hereditary factors play a certain part which cannot be precisely defined. Among siblings of ulcer patients, ulcers occur approximately twice as often, and among fathers, about five times as often as in persons of comparable age in the general population. Ulcus duodeni can be regarded as a relatively sharply defined feature which a person either has or does not have. But the individual contributing factors can possibly be found in pathogenic conditions which are quantitatively variable, such as can be detected in the acid production of the stomach or in the excretion of uropepsinogen. These factors could be dependent on numerous genes. The possibility suggests itself for ulcus duodeni, also, to conceive of the genetic component in the etiology as a special unfavorable configuration of several quantitatively variable factors. Here, again, sex plays an important part. In childhood, duodenal ulcers have approximately the same frequency in both sexes, but after puberty they become three to four times as frequent in the male as in the female sex, and only after the fifth decade of life does the ulcer incidence among women reach values comparable to that among men. These differences are so noticeably parallel to the normal sex differences in the acid production of the stomach that a causative relationship can be suspected. With puberty, the acid production of the stomach in the male sex increases distinctly, so that a distinct sex difference exists in adulthood. Certainly genetic influences are not the only determining factors in ulcers, however, as is demonstrated in particular by twin studies. Among 72 monozygotic pairs of twins with ulcers, only 18 were concordant, and among 84 dizygotic twins, 9. At least concordance is 2.5 times as frequent for monozygotic as for dizygotic twins.

Diabetes mellitus has been considered for a long time as a model example of a hereditary metabolic disorder. Dominant, recessive, and polygenic hereditary determination have been supported with equal emphasis. In contrast to most monomeric metabolic ailments, diabetes mellitus is not a congenital disorder, but one which appears only during the course of life and which cannot be demonstrated in advance even with special methods of examination.

A severe case of diabetes which requires insulin appears to be distinguished from the normal state with sufficient sharpness, and yet there are transitional states and doubtful cases. Especially in senile diabetes, which is only temporarily accompanied by excretion of sugar when a particular diet is used, and for the diagnosis of which one relies on the response to a double dose of glucose, the decision whether or not diabetes is present can be a difficult one. One then speaks of pre-diabetes. The boundary between the still normal and the pathological response to a double dose of glucose is drawn rather arbitrarily. Functionally, there is thus no sharp boundary between the normal condition and mild diabetes. From the anatomical point of view, it is still harder to draw the boundary line. The most characteristic morphological feature of diabetes mellitus is the decrease in the beta cells of the pancreatic islets. In this respect, however, there is a gradual transition to the norm (see Fig. 66). Among the parents of diabetic patients, in general 5–10 per cent are themselves diabetics, and among the siblings of patients, 3–9 per cent. The difference between the two groups of relatives presumably is due to the fact that the siblings often have not yet reached the age at which the disease manifests itself. In diabetes, also, environmental factors are important. The effect of overweight and the effect of repeated pregnancies are the best-known of these. Thirty-eight out of 67 pairs of monozygotic twins were found to be concordant, compared to only 10 out of 77 dizygotic pairs.

If one assumes that frequent diseases are determined by monomeric recessive genes, as has been discussed for diabetes and for the tendency toward "allergies," one arrives at a surprisingly high frequency of heterozygous carriers of the genes in the population. About 2–4 per cent of all people in the population of central and western Europe or of the United States of America become diabetic in the course of their lives, although usually only in old age. Twin studies show that not everyone who is genetically predisposed to diabetes actually becomes a diabetic. If one partner of a monozygotic pair of twins is diabetic, in almost half the cases the other does not become a diabetic, although he agrees with his twin not only in all hereditary traits but frequently also in living conditions. If even among the partners of diabetic monozygotic twins almost 50 per cent do not become diabetics, we can assume that, among random persons who are genetically disposed toward diabetes, a considerably higher percentage does not become diabetic. The number of individuals in the population who are genetically disposed toward diabetes can thus be estimated for good reason to be at least 8 per cent. If we now assume that diabetes mellitus has monomeric recessive inheritance, as is stated in most textbooks, a gene frequency of $\sqrt{0.08} = 0.283$ corresponds to this homozygote frequency of 0.08. The frequency of the normal allele would then amount to $1 - 0.283 = 0.717$, and the frequency of heterozygous carriers of the trait, to $2 \times 0.283 \times 0.717 = 0.406$. Thus it would be found that approximately 40 per cent of the population had the gene for diabetes in the heterozygous state. If a gene is so frequent, it is no longer easy to differentiate between recessive and dominant inheritance. With a high frequency of recessive genes, it will happen fairly often that homozygous patients find heterozygous marriage partners. In this case, however, it is to be expected that 50 per cent of the children are again homozygous. "Pseudodominant" inheritance would not be a rare exception in this case, but an entirely ordinary occurrence.

For *"allergies,"* which are still more frequent than diabetes, recessive inheritance is also still being discussed. In this case, the frequency of the heterozygous carriers of the trait would be still greater. Van Arsdel and Motulsky rightly point out that a polygenic system which determines the susceptibility to "allergies" could explain the familial accumulation just as well. It is not necessarily contrary to the hypothesis of recessive hereditary determination of a frequent disease that it requires the assumption of a very high gene frequency. Concerning sickle cell anemia and thalassemia we know that, under certain conditions, genes which have severe pathological effects in the homozygous state can attain a high frequency in the population. The non-uniform, variable character, fluctuating in "penetrance" and "expressivity," of diabetes mellitus or of the "allergies" is a much stronger argument against recessive hereditary determination, for the effect of a homozygous pair of genes generally is exceptionally constant. There are practically no recessive hereditary diseases with irregular penetrance or strong fluctuations in expressivity. Thus, if one assumes recessive inheritance for diabetes mellitus or for allergies, one must call on the diffuse concept of penetrance in order to obtain reasonably satisfactory agreement between theory and observation. Polygenic heredity seems more probable in this instance.

HUMAN GENETICS TERMINOLOGY

Explanation of the most important concepts

a priori method (of Bernstein-Lenz-Hogben): A statistical method in which the theoretically expected frequency is calculated for homozygous children of two heterozygous parents in sibships of differing size with at least one homozygous child, in order then to be compared with the empirically observed frequency. The a priori method is superior to Weinberg's sibling method, since it more completely exhausts the information contained in the material and therefore contains a smaller probable error (see p. 48).

Alleles: Genes which occupy homologous loci in homologous chromosomes are called alleles. Within the population, there is often a series of different ("multiple") alleles for a specific locus. An individual can never have more than two genes in the relationship of allelism to each other. Alleles regularly separate from each other during the reduction division. Of one pair of alleles of a parent, only one gene at a time can be passed on to a child (exception: chromosome aberrations). The designations "dominant" and "recessive" apply to the relationship of the effects of two alleles to each other.

Autosomes: Those chromosomes which are not sex chromosomes (heterosomes) are called autosomes. Man has twenty-two pairs of autosomes and two sex chromosomes (XX or XY).

Basic defect: A homogeneous defect of structural or chemical nature which forms the basis of the multiple symptoms of a hereditary disease is designated as the basic defect of the hereditary ailment under consideration. For genetic reasons, it is probable that a homogeneous basic defect is responsible for most monomeric hereditary syndromes.

Chiasmata: Before the first maturation division, sets of two longitudinally split homologous chromosomes in the tetrad stage (four-strand stage) lie next to each other. Here, crossing over (chiasmata) occurs between two non-sister chromatids, that is, between one-half of one chromosome and one-half of the other. The chiasmata are the morphological basis of crossing over. In the longer chromosomes, chiasmata are observed frequently.

Chromatids: The two halves into which each chromosome divides longitudinally before the reduction division are called chromatids. The halves which have come from one chromosome are called sister chromatids. The chromosome pictures obtained from cell cultures regularly show a longitudinal division into two chromatids, which are held together by the centromere (also called kinetochore).

Chromatin: The intensely stainable nuclear substance of the cell of which the chromosomes are composed. Chromatin consists of deoxyribonucleic acid and histone. One distinguishes between euchromatin, which loses its ability to be stained in the resting nucleus, and heterochromatin, which maintains its capacity for staining in the resting nucleus. In numerous species, the sex chromosomes are especially rich in heterochromatin. Presumably this is the basis of the possibility of distinguishing male from female cell nuclei by means of the "sex chromatin."

Chromatin-positive: The name used in abbreviated medical terminology for individuals whose cell nuclei contain sex chromatin. Those individuals are called chromatin-negative in whom the sex chromatin is lacking. In man, no exceptions have so far been published to the rule according to which all individuals with two or more X chromosomes are chromatin-positive, and all individuals with only one X chromosome are chromatin-negative.[1]

Chromosome: The chromosomes are the intensely stainable thread- or loop-shaped components of the cell nucleus. They are present in duplicate in the fertilized egg cell ("zygote") and in all body cells, as a "diploid" set of chromosomes. After the maturation division, only a single ("haploid") set of chromosomes is present in the germ cells. The chromosomes have the ability to duplicate themselves identically. Every chromosome has a centromere, to which a spindle fiber is attached during mitosis or meiosis in order to pull apart the daughter chromosomes (or, in meiosis, the homologous chromosomes) created by the identical duplication. Chromosomes and spindle apparatus normally serve for the equal distribution of the genetic material between the two daughter cells created by a mitosis. In meiosis, the diploid set of chromosomes is reduced to the haploid set of the mature germ cells (gametes).

In meiosis, homologous parts are exchanged between the homologous partners of a chromosome pair (crossing over).

For purposes of human genetics, it is sufficient to regard the chromosomes as linearly arranged groups of genes; the only known function of the group arrangement is the regular distribution of the genetic material to the daughter cells. Nothing definite is known in man about functional relationships of the genes lying together in one chromosome.

Chromosome aberration: Deviation from the normal number of chromosomes due to a disturbance of meiosis, which leads to pathological conditions.

Codominant (combinant): Refers to alleles whose effects are recognizable side by side, without one's being influenced by the other.

Consanguinity: Blood relationship, that is, possession of common ancestors in the preceding few generations of the ascending line.

Crossing over: The mechanism which leads to the exchange of chromosome segments between homologous chromosomes is called crossing over. Crossing over occurs before the first maturation division and always involves only one-half (chromatid) of the two homologous chromosomes which are paired in the tetrad stage and divided longitudinally. The chiasmata (*q.v.*) are the morphological basis of crossing over. Crossing over has the result that genes lying in one chromosome are not always passed on together to the descendants.

Deoxyribonucleic acid (DNA): Essential component of the cell nucleus which carries the genetic information. The DNA molecules are composed of nucleo-

---

[1] A case of a chromatin-positive XO individual was reported in July, 1960, at the International Congress of Endocrinology in Copenhagen.

tides in the form of a chain. Each nucleotide contains an organic base (adenine, guanine, thymine, cytosine), a sugar (deoxyribose), and phosphate. The specific nature of the genes is based on the specific sequence of base pairs in the DNA molecule.

Diploid: The double set of chromosomes of the body cells and of the germ cells before the maturation divisions is designated as diploid.

Dominant: Here, an originally strict definition, which is almost never applicable in human genetics, is distinguished from a looser term. Originally, a gene was called dominant if its effect was the same in the heterozygous as in the homozygous condition, i.e., if no effect of its allele was recognizable in the presence of the dominant gene. In this case, the allele is called recessive. Dominance and recessivity are complementary concepts. A gene is not dominant or recessive in itself, but only in its behavior with respect to a certain allele. In human genetics, the homozygous state of "dominant" anomalies is usually unknown. Here it is customary to speak of dominance if a gene has a distinctly recognizable pathological effect even in the heterozygous state, without considering whether this effect is equal to that of the homozygous state. "Dominant" genes in this sense presumably often have a more pronounced effect in the homozygous than in the heterozygous state. For the more specific characterization of the dominant gene effect, the following designations can be used:

Completely dominant applies to a gene whose heterozygous state cannot be distinguished from its homozygous state.

Conditionally dominant applies to a gene only the heterozygous state of which is known, this state being markedly different, however, from the homozygous state of the other allele.

Intermediately dominant applies to a gene which has a weaker effect in the heterozygous state than in the homozygous state (the expression has the same meaning as semidominant, partially dominant, or incompletely dominant).

Irregularly dominant applies to a gene whose heterozygous state is not always recognizable. The greater the irregularity of the gene effect, the more questionable it is whether this is really due to a "dominant" gene and not rather to a more complicated genetic mechanism.

Expressivity: Fluctuations in the degree of distinctness of a feature are explained by the "expressivity" of the responsible gene. This expression circumscribes a condition, without attempting an explanation. Low expressivity (low distinctness of the feature) passes continuously into missing penetrance (the feature is no longer recognizable).

Female carrier: A woman who is heterozygous for a recessive X-chromosomal gene. Half of the sons of a female carrier show the recessive X-chromosomal feature.

Gametes: The mature germ cells which have a haploid set of chromosomes. During fertilization, the two gametes produce the zygote, which has a diploid set of chromosomes.

Gene mutation: A mutation which involves only a single gene.

Gene product: The specific protein or polypeptide molecules which depend on the presence of a gene are called primary gene products. The primary gene products are frequently enzymes. Enzymes can synthesize most of the other molecules, such as complex polysaccharides, lipids, and even deoxyribonucleic acid and ribonucleic acid. However, enzymes cannot synthesize protein,

For this, ribonucleic acid is required, which translates the information contained in the genes into the synthesis of the gene products.

Genes: The elementary parts of the chromosomes, which cannot be separated by crossing over into smaller parts and which have a specific, homogeneous function. These two features of the definition do not exactly coincide in all cases. In general genetics, therefore, one differentiates between the non-crossing-over entity (recon) and the functional entity (cistron). Finally, one can still assume the smallest unit to be the muton, that is, the portion of the site of a gene which can be altered independent of a neighboring one by means of mutation. In human genetics, no error occurs in practice if the old gene concept is adhered to, which includes homogeneous transmission to the descendants, homogeneous function, and homogeneous mutability. The genes consist of deoxyribonucleic acid molecules, with whose specific composition of purine and pyrimidine bases the genetic information is given. Through the agency of ribonucleic acid, each gene produces a gene product, that is, a specific molecule or a part of a molecule. The genes are arranged linearly in the chromosomes. The place which is occupied by a gene is called its locus. The total number of genes in man is unknown; the estimates range between several thousand and over one hundred thousand.

Genome: The haploid set of chromosomes and the genes located in it.

Genome mutation: A mutation which alters the normal number of chromosomes of an individual. If the result of this is a whole multiple of the haploid set of chromosomes, one also speaks of ploidic mutation.

Genotype: The sum of the genes of an organism.

Haploid: A set of chromosomes in which only one chromosome of each homologous pair is present is called haploid. The gametes have a haploid set of chromosomes.

Hemizygous: A male individual is called hemizygous with respect to an X-chromosomal gene. Since only one X chromosome is present in the male sex, the homologous gene is missing. An individual is also called hemizygous with respect to a gene whose homologous partner is absent owing to the lack of a chromosome section. In man, this has not been demonstrated with certainty.

Heterogametic: The sex which has two different sex chromosomes, therefore, in man, the male.

Heterogeneous: A disease which actually is a group of like, or at least not definitely distinguishable, hereditary diseases which are due to different genes is called heterogeneous.

Heterozygous: An individual in whom the same gene locus on the two homologous chromosomes of a pair is occupied by two different alleles is called heterozygous with respect to this locus.

Holandric: Inheritance from the male to all male descendants. The Y chromosome, and thereby the male sex, is inherited by holandric heredity. Abnormal genes with holandric heredity are unknown.

Hologynic: Inheritance from a woman to all female descendants. Occurs in Drosophila as a consequence of a non-disjunction of the X chromosome. In man, hologynic heredity is not known.

Homogametic: The sex which has two like sex chromosomes, therefore, in man, the female.

Homogamy: If chance does not prevail in matings within a population, but if like genes are joined with higher-than-chance frequency through mating, this is called homogamy. The opposite, random mating, is called panmixia.

Human genetics: The study of human differences which are determined by heredity. Since it is impossible to see from the beginning whether or not the differences are determined by heredity, human genetics frequently concerns itself also with non-hereditary differences. The proof that differences determined by environment are not inherited is also a field of operation for human genetics. To this extent, human genetics can also be defined more generally as the science of the causes of individual differences.

Incomplete dominance: See Dominance.

Lethal factors: Disorders of the genetic material which lead to the death of the zygote, or of the individual developing from it, before attaining the age of reproduction. This can be a matter of lethal genes or deficiencies due to missing sections.

Lethal gene: A mutant gene which acts as a lethal factor.

Linkage: Non-allelic genes which lie in the same pair of chromosomes are called linked. If they are in the same chromosome ("coupling" phase), they are transmitted together to the descendants (exception: crossing over); if they lie in the two different chromosomes of a homologous pair, a child obtains either one or the other gene (exception: crossing over has the result that a child may receive both genes). Crossing over is so frequent that no great practical significance is attached to linkage.

Locus: The place in a chromosome which is occupied by a gene. Homologous loci can be occupied by alleles or by identical genes (heterozygous and homozygous state, respectively).

Maturation division: Meiosis.

Meiosis: The cell division by which, in two steps, the haploid gametes are produced from their diploid precursors.

Monogenic: Determined by a single gene locus.

Monomeric: Monogenic.

Monosomic: An individual who no longer has the complete diploid set of chromosomes owing to loss of a chromosome is called monosomic ($2n - 1$ instead of $2n$ chromosomes).

Mosaic: An individual composed of cells of different genotypes. Mosaics can come about by means of somatic mutation or by means of transplanting of cells from one individual to another, for example in dizygotic twins.

Multiple allelism: See Alleles.

Mutant: A mutant gene; sometimes used also for an individual having a mutant gene.

Mutation rate: The relative number of gametes with a new mutation, referred to the total number of gametes. Since each individual is created from two gametes, the number of cases of an illness due to a new mutation per person in the population is twice as large as the mutation rate.

Mutations: Discontinuous alterations of the genetic material which occur in part without known cause ("spontaneous") and in part as a result of the action of ionizing radiations or of certain chemical substances. Mutations can affect a gene (gene mutation), a chromosome section (chromosome mutation), or the number of chromosomes (genome mutation).

Non-disjunction: A disturbance of the normal course of meiosis in which two homologous chromosomes remain together. Non-disjunction is the cause of a genome mutation.

Panmixia: See Homogamy.

Partial dominance: See Dominant (intermediately dominant).

Penetrance: If the phenic effect of a gene is shown not regularly but only in a certain percentage of cases, one speaks of a penetrance of so-and-so-many per cent. If the penetrance of a gene deviates strongly from 100 per cent, doubts about the genetic hypothesis are justified.

Phene, phenic: The features which are assigned to certain genes or configurations of genes are called phenes. The expression is not equivalent to phenotype.

Phenocopy: If an exogenous influence leads to the same phenic characteristic as does a certain genotype, the first instance is called a phenocopy.

Phenogenetics: The science which attempts to explain the chain of causality between genotype and phenotype. Phenogenetics is more a part of developmental physiology than of genetics.

Phenotype: The manifest picture of an organism, that is, the totality of its external and internal characteristics, those which are genetically determined as well as those depending on the environment. The expression has frequently been misunderstood and usually can be dispensed with.

Plasma: The components of the cell other than the cell nucleus. It would be better always to speak of cytoplasm, since plasma, to the physician, means the blood fluid from which the corpuscular elements of the blood have been removed but in which, in contrast to serum, fibrinogen is still contained. For human genetics, the cytoplasm is without particular interest.

Pleiotropism: A gene which determines several characteristics is called pleiotropic or polyphenic.

Point mutation: A mutation which is not accompanied by a morphologically demonstrable change in the chromosome. In practice, it may be regarded as equivalent to a gene mutation.

Polygenic: A feature, usually a quantitatively variable characteristic, is polygenic if it is dependent on numerous genes. Most quantitatively variable physiological characteristics have polygenic inheritance. The involved genes are also designated as polygenes. Their specific effect is too small to be regularly recognizable.

Polymerism: Polygenic effect in which the participating genes are of equally strong effectiveness.

Polymorphism: The simultaneous occurrence of two or more different genotypes in the population, whose ratio is kept in equilibrium by means of selection. Example: The genes for sickle cell anemia and for normal hemoglobin in malaria regions.

Polyphenia: Pleiotropism.

Population genetics: The science of the genetic composition of the population. Population genetics seeks to determine gene frequencies and to detect the selective influences which determine the gene frequencies.

Proband: The starting case from which a genetic investigation is undertaken. For methodological reasons, it is necessary to characterize the probands, and the secondary cases detected through them, as such.

Pseudo-alleles: Closely linked genes which, in general, are transmitted as a unit, since because of the close linkage crossing over is observed very rarely. In man, pseudo-allelism has not so far been demonstrated. It is being discussed in connection with the Rh blood groups.

Recessive: See also Dominant. A gene is recessive in its behavior with respect to its allele if its effect is not recognizable in the presence of the latter. Only some of the genes which are usually designated as recessive correspond to this strict definition. It is customary to call all genes recessive which show a distinctly detectable effect only in the homozygous state, even if they also exhibit partial manifestations in the heterozygous state.

Reduction division: Meiosis.

Reproduction probability (effective fertility): The average number of children of patients with a specific hereditary disease (including patients who died early in life), in relation to the average number of children of comparable individuals who do not have the hereditary disease. The reproduction probability is a measure of the selection disadvantage of a hereditary disease. It may be reduced for biological, psychological, and sociological reasons.

Selection: The process which determines the relative share of the different genotypes in the propagation of the population. The selective effect of a gene can be defined by means of the reproduction probability of the carriers of the gene.

Semidominant: See Dominant.

Sex chromatin: Numerous somatic cell nuclei in the resting state contain accumulations of chromatin, which usually are attached to the nuclear membrane as sharply circumscribed corpuscles. Sex chromatin is found to be present only in individuals who have two X chromosomes, that is, normally only in the female sex. Double sex chromatin bodies in a cell point to the abnormal presence of three X chromosomes.[2]

Sex chromosomes: The chromosomes on which sex determination depends. Female individuals have two X chromosomes; males, one X and one Y chromosome. The X chromosome of man contains numerous genes which have nothing to do with sex determination. The Y chromosome is responsible for the development of the male gonads. Other Y-chromosomal genes have not been definitely established in man.

Sibling method: A statistical procedure for the correction of selection in family investigations which is based on the fact that one can detect only sets of siblings with at least one afflicted member. In this procedure, the siblings are counted starting from each proband, but the probands themselves are not included. The loss of information due to this makes the procedure inferior to the a priori method.

Somatic mutation: A mutation which involves a somatic cell and its daughter cells, that is, only a part of the organism. According to one hypothesis, some tumors are regarded as consequences of somatic mutation.

Specificity: A gene which can have qualitatively differentiated effects in different individuals has a low "specificity." Since this practically never occurs, the expression can be dispensed with. The high degree of specificity of gene effects is a matter of general experience in human genetics.

[2] See comment on page 182.

Tetrads: In the first meiotic division, the longitudinally divided (into two sister chromatids each) homologous chromosomes lie next to each other as "tetrads."

Translocation: Attachment of a chromosome fragment to a non-homologous chromosome. Translocation is a form of chromosome mutation.

Trisomy: Anomaly of the number of chromosomes, in which, in addition to the normal diploid set, one (simple trisomy) or several (double trisomy, etc.) chromosomes are present. Trisomic individuals come about through non-disjunction. In man, mongolism is known as an autosomal trisomy, Klinefelter's syndrome (XXY) and the XXX condition in the female as trisomy of the sex chromosomes.

Zygote: The fertilized egg cell from which the new individual develops.

MEDICAL GLOSSARY

Adrenaline: Hormone of the adrenal medulla and of the sympathetic nervous system.

Agammaglobulinemia: Hereditary (recessive X-chromosomal) or acquired anomaly in which the γ-globulin of the serum is diminished to a high degree.

Allergy: A condition of altered reactivity which is due to the fact that the body has formed antibodies against certain antigens which react with the latter. The expression "allergy" is often also applied in medicine to conditions which may occur as a consequence of an antigen-antibody reaction but for which it has not been proved that they always or usually are due to this mechanism. Conventionally, one includes primarily asthma bronchiale, eczema of infants, and hay fever among the "allergies."

Alveolar pyorrhea: Purulent inflammation of the supporting tissues of the teeth.

Amaurosis: Blindness.

Amino acids: The constituents of protein. Organic acids with an NH_2 group.

Anemia: A decrease in the red pigment per unit volume of blood.

Anemia, hemolytic: Anemia which is due to increased destruction of red blood corpuscles (hemolysis).

Anemia, hypochromic: Anemia in which the hemoglobin content of the red blood corpuscles is reduced.

Anemia, microcytic: Anemia in which the size of the red blood corpuscles is decreased.

Anisocytosis: Abnormal variability in the size of the red blood corpuscles.

Antibody: A protein substance formed by the body in response to a specific antigen, and reacting specifically with this antigen.

Antigens: Substances, usually of protein nature, against which the body can form specific antibodies; *see also* Allergy.

Antihemophilic globulin A and B: Two different fractions of the blood plasma which are necessary for the formation of thromboplastin, a substance which initiates the coagulation of blood.

Antiserum: A serum containing antibodies.

Aortic aneurysm: A pathological dilatation of the main artery.

Asthenic: Equivalent to frail. Asthenic habitus means a body shape with narrow thorax, poor muscular and fat development, and weak connective tissue.

Azoospermia: Absence of motile sperm in the semen.

Bilirubin: Bile pigment, decomposition product of the red blood pigment.

Blepharophimosis: Narrowing of the eyelid aperture.

Blood groups: Antigenic properties of the red blood corpuscles, which can lead to the formation of specific antibodies.

Bronchiectasis: Dilatation of the bronchi.

Cartilaginous exostoses: See Exostoses.

Cataract: Clouding of the lenses.

Chondrodystrophy: In the medical literature, the term "chondrodystrophy" is applied to a variety of skeletal anomalies. Here, it is equivalent to the more internationally used expression "achondroplasia," a dominant hereditary anomaly of the growth of the extremities with involvement of the skull (drawn-in root of the nose) and normal length of trunk.

Cochlea: The coil-shaped organ of acoustical perception in the inner ear.

Collagen fibers: The connective tissue fibers which are present in broad, non-branching bundles everywhere in the body and which have high tensile strength and little elasticity.

Constitution: Constitution sometimes means the genotype, but usually the phenotype to the extent that it gives the impression of a unified structure. The expression "constitution" is so vaguely defined that it should be avoided whenever possible.

Cretinism: Endemic stunted form of man which occurs in goiter regions, with a high degree of physical and mental developmental disorder, in which a functional disorder of the thyroid gland is most prominent.

Cubitus valgus: A deviation of the position of the lower arm with respect to the upper arm in the radial direction (side of the thumb).

Cyanosis: Bluish coloration, visible predominantly on the lips and on the ends of the extremities.

Dermal ridges: See Papillary pattern.

Dermatoglyphics: See Papillary pattern.

Deuteranomaly: Defective red-green vision with decreased sensitivity to green and displacement of the brightness maximum toward the long-wave side of the spectrum.

Deuteranopia: High degree of defective green vision.

Diabetes insipidus: Increased excretion of urine without sugar content (insipidus: tasteless). The cause lies in the diminished concentration capacity of the kidneys.

Drosophila: Different types of fruit flies which are especially easy to breed and therefore play a large role as objects of experiments in genetics.

Dysostosis cleidocranialis: A dominant hereditary anomaly predominantly of bones built upon connective tissue (skull, clavicles).

Ectodermal dysplasia: Different anomalies, due to predisposition, of the outer germinal layer (ectoderm) from which skin, hair, nails, and teeth develop are designated as ectodermal dysplasias.

Embryo: The human germinal product is designated as an embryo up to the beginning of the third month of pregnancy (embryo: "in moss," because the chorion is still covered on all sides with villi). When the later body shape is essentially formed, one no longer speaks of an embryo but of a fetus.

Embryopathy: An exogenous disease of the embryo, which leads to permanent damage (for example, rubella embryopathy).

Entoderm: The inner germinal layer from which the gastrointestinal tract and its appendicular glands (liver, pancreas, some endocrine glands) develop.

Enzyme: Ferment. A substance of protein nature which activates a specific chemical reaction.

Eosinophilic cells: White blood corpuscles whose cytoplasm contains granules which can be stained red with eosin.

Epicanthus: Fold of skin extending over the inner corner of the eye (canthus).

Epidermolysis bullosa: Anomalies of the skin in which the upper layer (epidermis) scales off in blisters (bullae).

Epiphyses: The ending and joint sections of the long bones which are formed from cartilage and separated from the central part by the growth cartilage.

Erythroblastosis: Flushing out of nucleated red blood corpuscles from the blood-forming centers into the peripheral blood.

Erythrocytes: The red blood corpuscles.

Erythropoiesis: The formation of red blood corpuscles.

Essential: In some ways equivalent to idiopathic. A disease is called essential if its real cause is unknown but if one wants to distinguish the disease from merely symptomatic forms. The symptomatic form is considered a secondary consequence of another basic illness. In contrast to this, the essential form is regarded as an independent disease.

Etiology: The study of causes of disease. While etiology concerns the primary causes, pathogenesis is understood to be the physiological connection of conditions ranging from the primary cause to the symptoms of the disease.

Exostoses: Bone excrescences, predominantly at the growth commissures of the long hollow bones.

Ferment: Enzyme.

Fibrinogen: Precursor of fibrin, the fibrous material in coagulated blood which gives the coagulated blood its mechanical properties.

Foam cells: Cells whose cytoplasm has a foamy appearance as a result of being filled with stored substances (usually of lipid nature).

Fundus: The posterior part of the eye.

γ-globulin: A fraction of the serum protein which is important especially for defense against infections.

Glaucoma: Condition of increased pressure in the eye.

Glomerula: The elements of the kidney in which fluid from the blood passes into the beginning portions of the urinary passages, while protein and blood corpuscles are held back.

Glucuronic acid: An organic acid, which has an important detoxification function. Many substances are coupled to glucuronic acid for the purpose of excretion.

Glycogen: The animal starch which is stored predominantly in the liver and which serves as a source of blood sugar.

Glycogenolysis: Liberation of glucose (grape sugar) from glycogen.

Glycolytic: Participating in glycolysis, that is, in the conversion of glucose into lactic acid, which proceeds without utilization of oxygen.

Gout: Disease with painful deposits of uric acid in the joints.

Hamartoma: Tissue malformation similar to a tumor but not a genuine tumor.

Haptoglobin: Certain protein fractions of the serum which combine with hemoglobin are designated as haptoglobins.

Hematuria: Excretion of red blood corpuscles in the urine.

Hemochromatosis: An irregularly dominant hereditary disease of iron metabolism, with pathological deposits of iron-containing pigment in numerous organs, primarily in liver, pancreas, and spleen.

Hemoglobin: The red blood pigment.

Hemoglobinuria: Excretion of red blood pigment in solution in the urine.

Hemophilia: Bleeder disease.

Hepatomegaly: Enlargement of the liver.

Hidrotic ectodermal dysplasia: In contrast to anhidrotic ectodermal dysplasia, in which the sweat glands are absent, a type with normal sweat glands is designated as "hidrotic."

Histo-incompatibility: Tissue intolerance with respect to transplanted tissue.

Hyper-aminoaciduria: Increased excretion of amino acids in the urine.

Hypercholesterolemia: Increase of cholesterol, a lipid, in the serum.

Hyperglycemia: Increased glucose content in the blood.

Hyperkeratosis: Thickening of the horny layer of the skin.

Hyperlipemia: Increase of neutral fat in the blood serum.

Hypertension: Increased blood pressure.

Hyperuricemia: Increase of uric acid in the serum.

Hypochromic: See Anemia.

Hypodontia: Congenital deficiency of several teeth.

Hypoglycemia: Decrease in blood sugar.

Hypophosphatemia: Decrease of inorganic phosphate in the serum.

Hypothyroidism: Diminished production of thyroid hormone.

Hypotrichosis: Decreased hair growth.

Ichthyosis hystrix: A generalized skin disease with bristle-like thickenings of the skin.

Icterus: Yellow coloring due to impregnation of the tissues with bilirubin.

Idiocy: The most severe degree of feeble-mindedness. The intelligence level of the second year of life is not exceeded.

Imbecility: Medium severity of feeble-mindedness, which makes profitable activity impossible. Intelligence level corresponds to the third to seventh year of life.

Infantile: Pertaining to a small child, to the age of infancy.

Interphalangeal synostosis: A bony fusion of the joints between the bones of the fingers.

Intersex: A person whose sex is unclear, or who has features of both sexes.

Ionizing radiations: The energetic, penetrating rays which can make positive ions from electrically neutral atoms by removal of electrons are collectively referred to as ionizing rays (γ rays, X rays, corpuscular radiations).

Isoagglutinins: Antibodies which are directed against antigenic properties of red blood corpuscles of the same species and which agglutinate them, that is, collect them into clumps.

Isoimmunization: The process of antibody formation as a reaction to antigens of the same species.

Isolate: Region of inbreeding. A portion of a population, within which marriages are concluded preferentially.

Klinefelter's syndrome: An anomaly due to abnormal constitution of sex chromosomes (XXY), in which one finds sterility because of azoospermia, small testicles with tubular hypoplasia, and frequently eunuchoid features as well as slight feeble-mindedness.

Leukemia: Disease with unlimited growth of white blood corpuscles.

Leukocytes: The white blood corpuscles.

Leukopenia: A decrease in the number of white blood corpuscles.

Lipids: Collective term for the fats and fatlike substances in the body.

Liver cirrhosis: Disease of the liver in which proliferation of connective tissue as well as destruction and new formation of liver cells are most prominent.

Lymphocytes: Cells with round nuclei which occur predominantly in the lymph nodes, in the spleen, and in the blood.

Meconium ileus: Closure of the intestine (ileus) by tough fecal matter of the newborn (meconium).

Medial necrosis: Degenerative changes of the elastic fibers in the middle layer of the aorta.

Melanoderma: Dark discoloration of the skin due to deposit of pigment.

Mesenchyme: Loose-meshed tissue proceeding from the middle germinal layer (the mesoderm). Connective tissue develops from the mesenchyme.

Mesoderm: The middle germinal layer.

Metacarpus: Bone of the middle part of the hand.

Metatarsus: Bone of the middle part of the foot.

Microcytic anemia: See Anemia.

Microphthalmia: Abnormal smallness of the eyeball.

Mongolism: A severe anomaly of physical and mental development due to trisomy of an autosome. The eyelid openings are narrow and are directed obliquely from a high outer edge to a low inner edge; the inner corner of the eyelid may be covered by an epicanthic fold. The nose is only slightly developed. The tongue is large and, in older patients, fissured. The joints are over-extensible, and the musculature is weak. Usually, the patients are imbecile, never normal mentally.

Multiple sclerosis: A localized disease of the central nervous system with highly variable clinical symptomatology.

Muscle relaxant: Medication which is given in the practice of anesthesia to relax the muscles.

Myotonic dystrophy: A dominant hereditary disease with muscular disorders, cataracts, hypogonadism, and deterioration of character.

Nephrocalcinosis: Deposit of calcium in the kidney tissues.

Nephrogenic diabetes insipidus: A recessive X-chromosomal anomaly with the symptomatology of diabetes insipidus which, however, in contrast to ordinary diabetes insipidus, cannot be affected by hormones from the posterior lobe of the pituitary gland.

Neurofibromatosis: A dominant hereditary disease in which one finds coffee-with-milk spots (café au lait) of the skin and polymorphous tumors (fibromas, neurinomas).

Neuropathy: A diffuse collective term for a great variety of behavioral disorders which are regarded as "constitutionally" caused.

Normal distribution: The Gaussian shape of the binomial distribution.

Nosology: The method of classifying diseases.

Nystagmus: Jerky twitching of the eye.

Oenothera: Evening primrose. With the evening primrose De Vries carried out his studies on mutations.

Oligospermia: Abnormally low number of sperms in the semen.

Opisthotonus: Increased state of tension of the extensor muscles of the neck and back.

Osteomalacia: Softening of the bones due to decrease of the mineral content of bone.

Osteoporosis: Decrease of the basic protein substance of bone.

Otosclerosis: Localized bone disease of the inner ear or of the auditory ossicles which can lead to deafness.

Oxalosis: A metabolic disorder, presumably of glycine metabolism, in which oxalates are excreted in abnormal amounts and deposited in different organs.

Papillary pattern (dermatoglyphics): The dermal ridge patterns of the finger tips and the palm (not to be confused with the main line relief of the hand).

Pathogenesis: See Etiology.

Pfaundler-Hurler disease: Dysostosis multiplex; gargoylism. Anomaly of the muco-polysaccharide metabolism which leads to skeletal disorders, feeble-mindedness, liver and spleen enlargement.

Phages: Types of virus which multiply in bacteria and destroy these in the process (bacteriophages).

Plasma: The blood fluid without the corpuscular components suspended in it.

Polycystic kidney disease: Disease in which the kidneys contain numerous cystic spaces.

Polydipsia: High intake of liquids.

Polyuria: Excretion of abnormally large amounts of urine.

Postaxial polydactyly: Having many fingers, with one (or more) additional finger(s) on the side of the little finger.

Progressive muscular dystrophy: A group of diseases with primary disturbance of muscle function.

Protanomaly: Defective red vision. The sensitivity to red is diminished. The brightness maximum is displaced toward the short-wave side of the spectrum.

Protanopia: High degree of defective red vision.

Pseudoglioma: Dense clouding behind the lens of the eye which cannot be distinguished externally from a tumor (glioma) is called pseudoglioma.

Pseudohermaphroditism: A condition in which gonads of one sex are found beside sex features of the other sex. Usually, the expression ps. is reserved for those cases in which features of both sexes can be recognized externally. Genuine hermaphroditism, on the other hand, is a condition in which female and male gonadal tissue is found at the same time.

Pseudopubertas praecox: Premature development of the secondary sex features without concurrent maturation of the sex glands.

Psychosis: Non-psychogenic, severe disturbance of psychic functions.

Psychosomatics: A direction in medicine which takes seriously especially the subjective aspect of illness and also etiologically traces back somatic symptoms of disease to subjective forces.

Ptosis: Inability to lift the upper eyelids.

Pyelonephritis: Inflammation of the kidneys affecting also the renal pelvis.

Renal diabetes: Excretion of sugar in the urine as a result of a disorder of the renal tubuli.

Retinitis pigmentosa: Degenerative illness of the retina with pigmentation.

Rhabdomyoma: A benign tumor of the transversely striated musculature.

Rickets: Disease of growing bone due to insufficient mineral deposits. The most frequent cause is lack of ultraviolet light.

Schizophrenia: Psychosis in which the lack of connectedness between affects and thoughts, and affective dullness, are most prominent.

Sclera: The dermis of the eye.

Scoliosis: Lateral curvature of the vertebral column.

Serum: The blood fluid remaining after the formation of coagulated blood.

Spastic hypertrophy of the pylorus: Disease of the first weeks of life, in which the outlet of the stomach is closed owing to hypertrophy of the circular musculature of the pylorus.

Spermatocytes: The stages of spermatogenesis proceeding from the spermatogonia. From one spermatocyte, four sperms are produced through the two maturation divisions.

Spherocytic anemia: A hemolytic anemia with erythrocytes of spherical shape.

Splenomegaly: Enlargement of the spleen.

Spongiosa: The marrow cavity of bone, with spongelike structure of the bone spicules, in contrast to the corticalis, the compact outer layer.

Syndactyly: Absence of separation of neighboring fingers or toes. In rare cases, also secondary fusion.

Syndrome: Originally, regularly recurring combinations of disease symptoms which cannot be considered as independent diseases were designated as syndromes. The meaning of the word "syndrome" has gradually shifted. Numerous anomalies designated as syndromes are defined more uniformly and sharply than are some diseases. An important reason for the fact that the etiological and pathogenic homogeneity of numerous syndromes is still insufficiently realized is the physician's lack of familiarity with genetics.

Tapetoretinal degeneration: Retinitis pigmentosa.

Telangiectasis: Dilatation of the smallest blood vessels.

Thalassemia: Microcytic anemia, which is widespread in the Mediterranean region. Thalassemia is a heterogeneous disease.

Thrombocytes: Blood platelets, which play an important part in coagulation.

Thrombopenia; thrombocytopenia: Diminished number of thrombocytes in the blood; leads to tendency to bleeding.

Thymus hyperplasia: A thymus gland which is distinctly recognizable in radiographs and which exceeds the central shadow on both sides, is found in about 30 per cent of healthy infants.

Tobacco mosaic disease: A virus disease of the tobacco plant in which the leaves have a mosaic-like checkered appearance.

Tuberous sclerosis: A dominant hereditary disease with differing forms of tissue malformations in the brain, the facial skin, and in internal organs, primarily heart and kidneys.

Tubuli of the kidneys: The collection tubules in the kidneys in which the chemical composition of the primary urine undergoes considerable changes because of back-resorption of water and different substances. *See also* Glomerula.

Tubuli of the testicles: The canaliculi of the testicles which are lined with the sperm-forming epithelium.

Ulcus duodeni: Ulcer of the duodenum.

Vaccinia virus: The virus of cowpox which serves for vaccination.

Xanthoma: Yellowish, node-shaped lipoid deposit in skin and hypodermis.

BIBLIOGRAPHY

ALDRICH, R. A., STEINBERG, A. G., and CAMPBELL, D. C.: Pedigree demonstrating a sex-linked recessive condition characterized by draining ears, eczematoid dermatitis and bloody diarrhea. Pediatrics **13**, 133–39, 1954.

ALLISON, A. C.: Notes on sickle cell polymorphism. Ann. Hum. Genet. **19**, 39–57, 1954.

———: Genetic control of human haptoglobin synthesis. Nature (Lond.) **183**, 1312–14, 1959a.

———: Metabolic polymorphisms in mammals. Amer. Nat. **93**, 5, 1959b.

———: Recent developments in the study of inherited anemias. Eugen. Quart. **6**, 155–66, 1959c.

ALLISON, A. C., and BLUMBERG, B. S.: Dominance and recessivity in medical genetics. Amer. J. Med. **25**, 933–41, 1958.

ATWOOD, K. C., and SCHEINBERG, S. L.: Somatic variation in human erythrocyte antigens. J. Cell. Comp. Physiol. **52**, Suppl. 1, 97–123, 1958.

BACHMANN, F.: Familienuntersuchungen beim kongenitalen Stuart-Prower-Faktor-Mangel. Arch. Julius-Klaus-Stiftg. **33**, 27–78, 1958.

BAILEY, N. T. J.: A classification of methods of ascertainment and analysis in estimating the frequencies of recessives in man. Ann. Eugen. (Lond.) **16**, 223–25, 1951a.

———: The estimation of the frequencies of recessives with incomplete multiple selection. *Ibid.* **16**, 215–22, 1951b.

BARTELS, E. D.: Heredity in Graves' disease. Op. ex domo biol. hered. hum. Univ. Hafniensis (Copenhagen), No. 2, 1941.

BASIC, M., and WEBER, D.: Über intrauterine Fruchtschädigung durch Röntgenstrahlen. Strahlentherapie **99**, 628, 1956.

BAUER, K. H.: Homoiotransplantation von Epidermis bei eineiigen Zwillingen. Beitr. Klin. Chir. **141**, 443–47, 1927.

BAUMANN, TH.: Die Mucoviscidosis als rezessives und irregulär dominantes Erbleiden. Eine klinische und genetische Studie. Basel and Stuttgart: Benno Schwabe & Co., 1958.

BEARN, A. G., and FRANKLIN, E. C.: Some genetical implications of physical studies of human haptoglobins. Science **128**, 596–97, 1958.

BECKER, P. E.: Dystrophia musculorum progressiva. Stuttgart: G. Thieme, 1953.

BECKER, P. E., and LENZ, F.: Zur Schätzung der Mutationsrate der Muskeldystrophien. Z. Menschl. Vererb. Konstitutionsl. **33**, 42–56, 1955.

BENZER, S., INGRAM, V. M., and LEHMANN, H.: Three varieties of human haemoglobin D. Nature (Lond.) **182**, 852–53, 1958.

BERARDINELLI, W.: An undiagnosed endocrinometabolic syndrome: report of 2 cases. J. Clin. Endocr. **14**, 193–204, 1954.

BERGANN, G., and WIEDEMANN, E.: Beobachtungen in vier Sippen mit Teleangiektasia hereditaria haemorrhagica (Oslersche Krankheit). Deutsch. Arch. Klin. Med. **202,** 26, 1955.

BERGMAN, S., REITALU, J., NOWAKOWSKI, H., and LENZ, W.: The chromosomes in two patients with Klinefelter syndrome. Ann. Hum. Genet. **24,** 81–88, 1960.

BERNSTEIN, F.: Über die Ermittlung und Prüfung von Genhypothesen aus Vererbungsbeobachtungen am Menschen und über die Unzulässigkeit der Weinbergschen Geschwistermethode als Korrektur der Auslesewirkung. Arch. Rassenbiol. **22,** 241–44, 1929.

BEUTLER, E.: The hemolytic effect of primaquine and related compounds. A review. Blood **14,** 103, 1959.

BEUTLER, E., ROBSON, M. J., and BUTTENWEISER, E.: The glutathione instability of drug sensitive red cells. J. Lab. Clin. Med. **49,** 84, 1957.

BIERICH, J.: Das adrenogenitale Syndrom im Kindesalter. Ergebn. Inn. Med. Kinderheilk. N.F. **9,** 510–85, 1958.

BIGGS, R., and MACFARLANE, R. G.: Human blood coagulation and its disorders. 2d ed. Oxford: Blackwell Scientific Publications, 1957.

———: Haemophilia and related conditions: a survey of 187 cases. Brit. J. Haemat. **4,** 1–27, 1958.

BLANK, C. E.: Apert's syndrome (a type of acrocephalosyndactyly). Ann. Hum. Genet. **24,** 151–64, 1960.

BOHN, H., KOCH, E., KOCH, F., RICK, W., and RAU, R.: Die Erwachsenen-Mucoviscidosis als überaus häufige dominante erbliche Krankheit. Die Medizinische, pp. 1139–49, 1959.

BOIVIN, P., BOUSSER, J., ROBINEAUX, R., and PRINGUET, C.: Pancytopenie avec malformations multiples (Syndrome de Fanconi). Présentation de 3 nouveaux cas et revue de la littérature. Arch. Franc. Pediat. **15,** 1289, 1958.

BOLDRINI, M.: La fertilità dei biotipi. Saggio de demografia costituzionalistica. Milan: Soc. Ed. "Vita e Pensiero," 1931.

BÖÖK, J. A.: A contribution to the genetics of congenital clubfoot. Hereditas **34,** 289, 1948*a.*

———: The frequency of cousin marriage in three North Swedish parishes. *Ibid.* **34,** 252–55, 1948*b.*

———: A clinical and genetical study of disturbed skeletal growth (chondrohypoplasia). *Ibid.* **36,** 161–80, 1950.

———: Genetical investigation in a North Swedish population. The offspring of first-cousin marriages. Ann. Hum. Genet. **21,** 191–223, 1957.

BÖÖK, J. A., FRACCARO, M., and LINDSTEN, J.: Cytogenetical observations in mongolism. Acta Paediat. **48,** 453–68, 1959.

BÖÖK, J. A., and SANTESSON, B.: Malformation syndrome in man associated with triploidy (69 chromosomes). Lancet **1,** 858–59, 1960.

BORBERG, A.: Clinical and genetic investigations into tuberous sclerosis and Recklinghausen's neurofibromatosis. Contribution to elucidation of interrelationship and eugenics of the syndromes. Op. ex domo biol. hered. hum. Univ. Hafniensis (Copenhagen), No. 23, 1951.

BOWLES, G. T.: New types of old Americans at Harvard and at eastern women's colleges. Cambridge, Mass.: Harvard University Press, 1932.

BRAESTRUP, C. B.: Past and present radiation exposure to radiologists from the point of view of life expectancy. Amer. J. Roentgenol. **78**, 988–92, 1957.

BREAKEY, V. K. St. G., GIBSON, Q. H., and HARRISON, D. C.: Familial idiopathic methaemoglobinaemia. Lancet **1**, 935–38, 1951.

BÜCHNER, F.: Von den Ursachen der Missbildungen und Missbildungskrankheiten. Münch. Med. Wschr. **97**, 1673–77, 1955.

BURKS, BARBARA: The relative influence of nature and nurture upon mental development. 27th Yearbook, National Society for the Study of Education **1**, 219, 1928.

BURT, C.: The inheritance of mental ability. Amer. Psychologist **13**, 1–15, 1958.

CANNON, J. F.: Hereditary multiple exostoses. Amer. J. Hum. Genet. **6**, 419, 1954.

CAPOTORTI, L., GADDINI DE BENEDETTI, R., and RIZZO, P.: Contributo allo studio dell' ereditarietà della S. di Marfan. Descrizione di un albero genealogico di quattro generazioni con un matrimonio fra consanguinei affetti. Acta Genet. Med. (Rome) **8**, 455–82, 1959.

CARLSON, E. A.: The bearing of a complex-locus in Drosophila on the interpretation of the Rh-series. Amer. J. Hum. Genet. **10**, 465–73, 1958.

CARTER, C., and SIMPKISS, M.: The "carrier" state in nephrogenic diabetes insipidus. Lancet **2**, 1069–73, 1956.

CARTER, C. O., and HARRIS, H.: Experimental and human genetics. *In:* Modern trends in pediatrics, pp. 16–40. London: Butterworth, 1958.

CARTER, T. C.: Recessive lethal mutation induced in the mouse by chronic γ-irradiation. Proc. Roy. Soc. [Biol.] **147**, 402–11, 1957.

———: Radiation induced gene mutation in adult female and foetal male mice. Brit. J. Radiol. **31**, 407–11, 1958.

CATHALA, J., DEMASSIEUX, M., and GEORGES-JANET, L.: Trois cas de polydysplasie ectodermique héréditaire chez le nourrisson. Bull. Soc. Méd. Hôp. (Paris) **73**, 1056–61, 1957.

CEPPELINI, R., DUNN, C. L., and INNELLA, F.: Immunogenetica II. Analisi genetica formale dei caratteri Lewis con particolare riguardo alla natura epistatica della specificità serologica Le[b]. Folia Hered. Pathol. **8**, 261–96, 1959.

CHATTERJEA, J. B., SWARUP, S., GHOSH, S. K., and RAY, R. N.: Hb. S-thalassemia disease in India. J. Indian Med. Ass. **30**, 4–8, 1958.

CHEESEMAN, E. A., KILPATRICK, S. J., STEVENSON, A. C., and SMITH, C. A. B.: The sex ratio of mutation rates of sex-linked recessive genes in man with particular reference to Duchenne type of muscular dystrophy. Ann. Hum. Genet. **22**, 235–43, 1958.

CHILDS, B., GRUMBACH, M. M., and VAN WYK, J. J.: Virilizing adrenal hyperplasia; a genetic and hormonal study. J. Clin. Invest. **35**, 213–22, 1956.

CHILDS, B., and SIDBURY, J. B.: A survey of genetics as it applies to problems in medicine. Pediatrics **20**, 177–218, 1957.

CHILDS, B., SIDBURY, J. B., and MIGEON, C. J.: Glucuronic acid conjugation by patients with familial nonhemolytic jaundice and their relatives. Pediatrics **23**, 903–13, 1959.

CHILDS, B., ZINKHAM, W., BROWNE, E. A., KIMBRO, E. L., and TORBERT, J. V.: A genetic study of a defect in glutathione metabolism of the erythrocyte. Bull. Johns Hopkins Hosp. **102**, 21–37, 1958.

CHU, E. H. Y., and GILES, N. H.: Human chromosome complements in normal somatic cells in culture. Amer. J. Hum. Genet. **11**, 63–79, 1959.

CLARK, P. J.: The heritability of certain anthropometric characters as ascertained from measurements of twins. Amer. J. Hum. Genet. **8**, 49–54, 1956.

CLARKE, C. A., EVANS, D. A. P., McCONNELL, R. B., and SHEPPARD, P. M.: Secretion of blood group antigens and peptic ulcer. Brit. Med. J. **1**, 603–7, 1959.

COCKAYNE, E. A.: Epicanthus and bilateral ptosis. Proc. Roy. Soc. Med. **24**, 847, 1931.

COLLEY, J. R., MILLER, D. L., HUTT, M. S. R., WALLACE, H. J., and WARDENER, H. E.: The renal lesion in angiokeratoma corporis diffusum. Brit. Med. J. **1**, 1266–68, 1958.

COURT-BROWN, W. M.: Nuclear and allied radiations and the incidence of leukemia in man. Brit. Med. Bull. **14**, 168–73, 1958.

CROCKER, A. C., and FARBER, S.: Niemann-Pick disease: a review of eighteen patients. Medicine (Balt.) **37**, 1–95, 1958.

CROW, J. F.: A comparison of fetal and infant death rates in the progeny of radiologists and pathologists. Amer. J. Roentgenol. **73**, 467–71, 1955.

CROWE, F. W., SCHULL, W. J., and NEEL, J. V.: A clinical, pathological, and genetic study of multiple neurofibromatosis. Springfield, Ill.: C. C Thomas, 1956.

CURTIUS, F.: Neuere Ergebnisse der erbbiologischen multiplen Sklerose-Forschung. Fortschr. Neurol. Psychiat. **27**, 161–84, 1959.

CZELLITZER, A.: Augenfehler. *In:* Handwörterbuch der sozialen Hygiene. Edited by GROTJAHN and KAUP. Leipzig: Vogel, 1912.

DAHLBERG, G.: Twin births and twins from a hereditary point of view. Stockholm: Bokförlags, A.-B. Tidens Tryckeri, 1926.

———: Mathematische Erblichkeitsanalyse von Populationen. Uppsala: Almqvist & Wiksell, 1943.

DALGAARD, O. Z.: Bilateral polycystic disease of the kidneys. A follow-up of two hundred and eighty-four patients and their families. Op. ex domo biol. hered. hum. Univ. Hafniensis (Copenhagen), No. 38, 1957.

DARWIN, G. H.: Marriages between first cousins in England and their effects. J. Statist. Soc. **38**, 153–82, 1875.

DAVENPORT, G. C., and DAVENPORT, C. B.: Heredity of skin pigmentation in man. Amer. Nat. **44**, 641–72, 1910.

DEBRÉ, R., DREYFUS, J.-CL., FRÉZAL, J., LABIE, D., LAMY, M., MAROTEAUX, P., SCHAPIRA, F., and SCHAPIRA, G.: Genetics of haemochromatosis. Ann. Hum. Genet. **23**, 16–30, 1958.

DENCKER, S. J.: A follow-up study of 128 closed head injuries in twins using co-twins as controls. Acta Psychiat. Scand., Suppl. 123, **33**, 1–125, 1958.

DENT, C. E., and HARRIS, H.: Hereditary forms of rickets and osteomalacia. J. Bone Joint Surg. [Brit.], **38 B**, 204–26, 1956.

DENYS, P., CORBEEL, L., EGGERMONT, E., and MALBRAIN, H.: Le syndrome de Lowe. Étude de la fonction tubulaire. Pédiatrie **13**, 639–60, 1958.

DIEHL, K., and VERSCHUER, O. von: Erbeinfluss bei der Tuberkulose. Jena: G. Fischer, 1936.

DOLL, R., and JONES, A.: Occupational factors in the aetiology of gastric and duodenal ulcers. Med. Res. Counc. Spec. Rep. (Lond.), No. 276, 1951.

DOUGLAS, A. S., and COOK, I. A.: Deficiency of antihaemophylic globulin in heterozygous haemophilic females. Lancet **2**, 616–19, 1957.

DOUGLAS, W. F., SCHONHOLTZ, G. J., and GEPPER, L. J.: Chondroectodermal dysplasia (Ellis–van Creveld syndrome). J. Dis. Childr. 97, 473–78, 1959.

DRINKWATER, H.: Phalangeal synostosis anarthrosis (synostosis ankylosis) transmitted through fourteen generations. Proc. Roy. Soc. Med. 10, Sect. Pathol., 60–68, 1917.

EBBING, H. C.: Über Diabetes insipidus, insbesondere seine erblichen Formen. Z. Menschl. Vererb. Konstitutionsl. 33, 415–24, 1956.

EDMUND, J.: Blepharophimosis congenita. Acta Genet. 7, 279–84, 1956.

EDWARDS, J. H., HARNDEN, D. G., CAMERON, A. H., CROSSE, V. M., and WOLFF, O. H.: A new trisomic syndrome. Lancet 1, 787–89, 1960.

EGENTER, A.: Über den Grad der Inzucht in einer Schwyzer Berggemeinde und die damit zusammenhängende Häufung rezessiver Erbschäden. Arch. Julius-Klaus-Stiftg. 9, 365–406, 1934.

EUGSTER, J.: Zur Erblichkeitsfrage des endemischen Kropfes. III. Die Zwillingsstruma. Untersuchungen an 520 Zwillingspaaren. Arch. Julius-Klaus-Stiftg. 11, 369–539, 1936.

FINNEY, D. J.: The truncated binomial distribution. Ann. Eugen. (Lond.) 14, 319–28, 1947–49.

FISCHER, J., LANDBECK, G., and LENZ, W.: Vergleichende Bestimmungen des antihaemophilen Globulins bei beiden Geschlechtern. Klin. Wschr. 36, 20–22, 1958.

FISHER, R. A.: The fitting of gene frequencies to data on rhesus reactions. Ann. Eugen. (Lond.) 13, 150, 1946.

———: The Rhesus Factor. A study in scientific method. Amer. Scientist 35, 95–103, 1947.

FLAMM, H.: Die pränatalen Infektionen des Menschen unter besonderer Berücksichtigung von Pathogenese und Immunologie. Stuttgart: G. Thieme, 1959.

FORD, C. E., and HAMERTON, J. L.: The chromosomes of man. Nature (Lond.) 178, 1020–23, 1956.

FORD, C. E., JONES, K. W., MILLER, O. J., MITTWOCH, U., PENROSE, L. S., RIDLER, M., and SHAPIRO, A.: The chromosomes in a patient showing both mongolism and the Klinefelter syndrome. Lancet 1, 709–10, 1959.

FORD, C. E., JONES, K. W., POLANI, P. E., DE ALMEIDA, J. C., and BRIGGS, J. H.: A sex chromosome anomaly in a case of gonadal dysgenesis (Turner syndrome). Lancet 1, 711–13, 1959.

FORD, C. E., POLANI, P. E., BRIGGS, J. H., and BISHOP, P. M. F.: A presumptive human XXY/XX mosaic. Nature (Lond.) 183, 1030–32, 1959.

FORSSMAN, H.: Two different mutations of the X-chromosome causing diabetes insipidus. Amer. J. Hum. Genet. 7, 21–27, 1955.

FOX, S. R.: Benign familial icterus: a report of three cases. Ann. Intern. Med. 50, 115–21, 1959.

FRACCARO, M., KAIJSER, K., and LINDSTEN, J.: Chromosome complement in gonadal dysgenesis (Turner's syndrome). Lancet 1, 886, 1959.

FRANCESCHETTI, A., BAMATTER, F., and KLEIN, D.: Valeur des tests cliniques et sérologiques en vue de l'identification de deux jumeaux univitellins dont l'un a été échangé par erreur. Bull. Acad. Suisse Sci. Méd. 4, 433–44, 1948.

FRANÇOIS, J., and DEWEER, J. P.: Albinisme oculaire lié au sexe et altérations caractéristiques du fond d'œil chez les femmes hétérozygotes. Bull. Soc. Belg. Ophtal. 102, 724–39, 1952.

FRANÇOIS, R., DREYFUS, J. D., MOURIQUAND, CL., BERTRAND, J., and RUITON-UGLI-
ENGO: Polycorie glycogénique du foie chez deux frères par insuffisance en
glucose-6-phosphatase. Sem. Hôp. (Paris) **33**, 11, 1957.

FRASER, F. C.: Medical genetics in pediatrics. J. Pediat. **44**, 85–103, 1954.

———: Causes of congenital malformations in human beings. J. Chron. Dis. **10**,
97–110, 1959.

FRASER, F. C., and SCRIVER, J. B.: A hereditary factor in chondrodystrophia cal-
cificans congenita. New Engl. J. Med. **250**, 272–77, 1954.

FRASER ROBERTS, J. A.: An introduction to medical genetics. 2d ed. London:
Oxford University Press, 1959.

FREIRE-MAIA, N.: Inbreeding levels in different countries. Eugen. News **4**, 127–38,
1957.

FROESCH, E. R., PRADER, A., LABHART, A., STUBER, H. W., and WOLF, H. P.: Die
hereditäre Fruktoseintoleranz, eine bisher nicht bekannte kongenitale Stoff-
wechselstörung. Schweiz. Med. Wschr. **87**, 1168–71, 1957.

GALTON, F.: Inquiries into human faculty. London: J. M. Dent & Sons, 1883.

GATES, R. R., and BHADURI, P. N.: The inheritance of hairy ear rims. Mankind
Monographs, **1**. Edinburgh: Mankind Quarterly, 1961.

GEDDA, L.: Studio dei Gemelli. Rome: Orizzonte Medico, 1951.

GERALD, P. S., and GEORGE, P.: Second spectroscopically abnormal methemoglobin
associated with hereditary cyanosis. Science **129**, 393, 1959.

GESELL, A., and THOMPSON, H.: Learning and growth in identical infant twins.
An experimental study by the method of co-twin control. Genet. Psychol.
Monogr. **6**, 1–124, 1929.

GLASS, B.: Discussion. Cold Spring Harbor Symp. on Quant. Biol. **11**, 22, 1950.

GOLD, J. J., and FRANK, R.: The borderline adrenogenital syndrome: an intermediate entity. Amer. J. Obstetr. Gynec. **75**, 1034–42, 1958.

GOLDSCHMIDT, R.: Theoretical genetics. Berkeley: University of California Press,
1955.

GOOD, R. A., and ZAK, S. J.: Disturbances in gamma globulin synthesis as "experiments of nature." Pediatrics **18**, 109–49, 1956.

GRAHAM, J. B., BARROW, E. M., and HOUGIE, C.: Stuart clotting defect. II. Genetic aspects of a "new" hemorrhagic state. J. Clin. Invest. **36**, 497–503, 1957.

GREBE, H.: Chondrodysplasie. Rome: Istituto Gregorio Mendel, 1955.

GROSS, R. T., HURWITZ, R. E., and MARKS, P. A.: An hereditary enzymatic defect
in erythrocyte metabolism: glucose-6-phosphate dehydrogenase deficiency.
J. Clin. Invest. **37**, 1176–84, 1958.

GROUCHY, J. DE: L'hérédité moléculaire. Conditions normales et pathologiques.
Rome: Istituto Gregorio Mendel, 1958.

HADORN, ERNST: Letalfaktoren in ihrer Bedeutung für Erbpathologie und Gen-
physiologie der Entwicklung. Stuttgart: G. Thieme, 1955.

HALDANE, J. B. S.: A method for investigating recessive characters in man.
J. Genet. **25**, 251–55, 1932.

———: A search for incomplete sex linkage in man. Ann. Eugen. (Lond.) **7**, 28–57,
1936.

———: The estimation of the frequency of recessive characters in man. *Ibid*. **8**,
255–62, 1937.

HAMILTON, M., PICKERING, G. W., FRASER ROBERTS, J. A., and SOWRY, G. S. C.:
The aetiology of essential hypertension. Clin. Sci. **13**, 11, 37, 267, 273, 1954.

HANHART, E.: Mongoloide Idiotie bei Mutter und zwei Kindern aus Inzesten. Acta Genet. Med. (Rome) **9,** 112–30, 1960.

HANSSON, H., LINELL, F., NILSSON, L. R., SÖDERHJELM, L., and UNDRITZ, E.: Die Chediak-Steinbrinck-Anomalie resp. erblich-konstitutionelle Riesengranulation (Granulagiganten) der Leukozyten in Nordschweden. Fol. Haemat. (Frankf.) **3,** 152–96, 1959.

HARDISTY, R. M.: Christmas disease in women. Brit. Med. J. **1,** 1039–1040, 1957.

HARNACK, G. A. VON, HORST, W., LENZ, W., and ZUKSCHWERDT, L.: Homotransplantation von Schilddrüsengewebe bei einem eineiigen Zwillingspaar. Deutsch. Med. Wschr. **83,** 549–55, 1958.

HARRIS, H.: A sex-limiting modifying gene in diaphysical aclasis (multiple exostoses). Ann. Eugen. (Lond.) **14,** 165–70, 1947–49.

HARRIS, H., and ROBSON, E. B.: Cystinuria. Amer. J. Med. **23,** 774, 1957.

HARTMANN, G., and THOENEN, H.: Zur Pathogenese der essentiellen Hyperlipämie. Schweiz. Med. Wschr. **89,** 1102, 1959.

HAUGE, M., and HARVALD, B.: Heredity in gout and hyperuricemia. Acta Med. Scand. **152,** 247–57, 1955.

HAVERKAMP BEGEMANN, N., and VAN LOOKEREN CAMPAGNE, A.: Homozygous form of Pelger-Huet's nuclear anomaly in man. Acta Haemat. **7,** 295, 1952.

HEBERER, F., KURTH, G., and SCHWIDETZKY-ROESING, I.: Anthropologie. Frankfurt: Fischer-Bücherei, 1959.

HELMBOLD, W.: Über den möglichen Aufbau des Rh-Genkomplexes. Blut **5,** 141–48, 1959.

HERNDON, C. M.: Genetics of the lipidoses. Proc. Ass. Res. Nervous Mental Dis. **33,** 239–58, 1954.

HILL, R. L., and SCHWARTZ, H. C.: A chemical abnormality in haemoglobin G. Nature (Lond.) **184,** 641–42, 1959.

HIRSCHHORN, K., and WILKINSON, CH. F.: The mode of inheritance in essential familial hypercholesterolemia. Amer. J. Med. **26,** 60–67, 1959.

HITZIG, W. H., and ZOLLINGER, W.: Kongenitaler Faktor-VII-Mangel. Familienuntersuchung und physiologische Studien über den Faktor VII. Helv. Paediat. Acta **13,** 189–203, 1958.

HOGBEN, L.: Nature and nurture. London: Allen & Unwin, 1935.

HOLLMANN, S.: Zur Biochemie der essentiellen Pentosurie, kongenitalen Galaktosämie und Phenylketonurie. Klin. Wschr. **37,** 737–42, 1959.

HOLUB, D. A., GRUMBACH, M. M., and JAILER, J. W.: Seminiferous tubule dysgenesis (Klinefelter's syndrome) in identical twins. J. Clin. Endocr. **18,** 1359–68, 1958.

HOLZEL, A., SCHWARZ, V., and SUTCLIFFE, K. W.: Defective lactose absorption causing malnutrition in infancy. Lancet **1,** 1126–28, 1959.

HOWELLS, W. W.: Correlations of brothers in factor scores. Amer. J. Phys. Anthrop. **11,** 121–40, 1953.

HSIA, D. Y.-Y.: Inborn errors of metabolism. Chicago: Year Book Publishers, 1959.

HSIA, D. Y.-Y., DRISCOLL, K. W., TROLL, W., and KNOX, W. E.: Detection by phenylalanine tolerance tests of heterozygous carriers of phenylketonuria. Nature (Lond.) **178,** 1239, 1956.

HSIA, D. Y.-Y., and GAWRONSKA KOT, E.: Detection of heterozygous carriers in glycogen storage disease of the liver (von Gierke's disease). Nature (Lond.) **183,** 1331–32, 1959.

HSIA, D. Y.-Y., NAYLOR, J., and BIGLER, J. A.: Gaucher's disease. Report of two cases in father and son and review of the literature. New Engl. J. Med. **261,** 164–69, 1959.

HUANG, I., HUGH-JONES, K., and HSIA, D. Y.-Y.: Studies on the heterozygous carrier in galactosemia. J. Lab. Clin. Med. **54,** 585–92, 1959.

HUNT, J. A., and INGRAM, V. M.: Allelomorphism and the chemical differences of the human haemoglobins A, S and C. Nature (Lond.) **181,** 1062–63, 1958.

HUNTLEY, C. C., and DEES, S. C.: Eczema associated with thrombocytopenic purpura and purulent otitis media. Report of five fatal cases. Pediatrics **19,** 351–61, 1957.

INGALLS, TH. H., AVIS, F. R., CURLEY, FR. J., and TEMIN, H. M.: Genetic determinants of hypoxia-induced congenital anomalies. J. Hered. **44,** 185–94, 1953.

INGRAM, V. M.: Gene mutation in human haemoglobin: the chemical difference between normal and sickle cell haemoglobin. Nature (Lond.) **180,** 326–28, 1957.

———: Constituents of human haemoglobin. Separation of the peptide chains of human globin. *Ibid.* **183,** 1795–1800, 1959.

———: Chemistry of the abnormal haemoglobins. Brit. Med. Bull. **15,** 27, 1959.

ISRAELS, M. C. G., LEMPERT, H., and GILBERTSON, E.: Haemophilia in the female, Lancet **1,** 1375–80, 1951.

ITANO, H. A., and NEEL, J. V.: A new inherited abnormality of human hemoglobin. Proc. Nat. Acad. Sci. USA **36,** 613–17, 1950.

ITANO, H. A., and ROBINSON, E.: Formation of normal and doubly abnormal haemoglobin by recombination of haemoglobin I with S and C. Nature (Lond.) **183,** 1959.

JACKSON, A. D. M., and FISCH, L.: Deafness following maternal rubella. Lancet **2,** 1241–44, 1958.

JOHNSON, S. A. M., and FALLS, H. F.: Ehlers-Danlos syndrome; a clinical and genetic study. Arch. Derm. Syph. **60,** 82–103, 1949.

JUNGCK, E. C., BROWN, N. H., and CARMONA, N.: Constitutional precocious puberty in the male. J. Dis. Childr. **91,** 138–43, 1956.

KAELIN, A.: Estimation statistique de la fréquence des tarés en génétique humaine. J. Genet. Hum. **7,** 67–91, 121–42, 243–95, 1958.

KAHLER, O.-H.: Beitrag zur Erbpathologie der Dysostosis cleidocranialis. Z. menschl. Vererb. Konstitutionsl. **23,** 216–34, 1939.

KALLMANN, F. J., and JARVIK, L. F.: Twin data on genetic variations in resistance to tuberculosis. *In:* Genetica della Tubercolosi e dei Tumori. Atti Simpos. Internaz. Torino 1957. Analecta Genetica **6.** Rome: Istituto Gregorio Mendel, 1958.

KALOW, W., and STARON, N.: On distribution and inheritance of atypical forms of human serum cholinesterase, as indicated by dibucaine numbers. Canad. J. Biochem. Physiol. **35,** 1305–20, 1957.

KAPLAN, I. I.: Genetic effects in children and grandchildren of women treated for infertility and sterility by roentgen therapy. Report of a study of thirty-three years. Radiology **72,** 518–21, 1959.

KEIZER, D. P. R.: Puberté précoce familiale. Arch. Franc. Pediat. **13,** 986–92, 1956.

KHERUMIAN, R., and PICKFORD, R. W.: Hérédité et fréquence des dyschromatopsies. Paris: Vigot Frères, 1959.

KINGSLEY, C. S.: Familial Factor V deficiency: the pattern of heredity. Quart. J. Med., N.S. **23,** 323–29, 1954.

KIRMAN, B. H., BLACK, J. A., WILKINSON, R. H., and EVANS, P. R.: Familial pitressin resistant diabetes insipidus with mental defect. Arch. Dis. Child. **31,** 59–66, 1956.

KIVALO, A., and KIVALO, E.: Juvenile amaurotic idiocy. Vacuolisation of lymphocytes in the healthy members of families involved. Ann. Paediat. Fenn. **4,** 191–95, 1958.

KLEIN, D., and FRANCESCHETTI, A.: Le dépistage des conducteurs de gènes pathologiques. Rev. Med. Suisse Rom. **79,** 369–99, 1959.

KLEIN, D., and KTÉNIDÈS, M.-A.: Au sujet de l'hérédité de l'idiotie amaurotique infantile (Tay-Sachs). J. Genet. Hum. **3,** 184–202, 1954.

KNOX, W. E., and MESSINGER, E. C.: The detection of the metabolic effect of the recessive gene for phenylketonuria. Amer. J. Hum. Genet. **10,** 53–60, 1958.

KOMAI, T.: Pedigrees of hereditary diseases and abnormalities found in the Japanese race. Kyoto, 1934.

KOMROWER, G. M., SCHWARZ, U., HOLZEL, A., and GOLBERG, L.: A clinical and biochemical study of galactosemia. Arch. Dis. Child. **31,** 254–64, 1956.

KOZINN, P. J., WIENER, H., and COHEN, P.: Infantile amaurotic family idiocy. A genetic approach. J. Pediat. **51,** 58–64, 1957.

KRAG, C. L.: The prevalence of diabetes mellitus among the aged tested in a community diabetes program. J. Geront. **8,** 325–27, 1953.

KUO, P. T., WHEREAT, A. F., and HORWITZ, O.: The effect of lipemia upon coronary and peripheral arterial circulation in patients with essential hyperlipemia. Amer. J. Med. **26,** 68–75, 1959.

LAMY, M., AUSSANNAIRE, M., JAMMET, M. L., and CARAMANIAN, M.: Cystinose avec syndrome de Toni-Debré-Fanconi. Étude clinique et biologique. Arch. Franc. Pediat. **11,** 806, 1954.

LAMY, M., and FRÉZAL, J.: La dysplasie chondroectodermique (Ellis van Creveld). *In:* L. Gedda (ed.): Novant' anni delle Leggi Mendeliane. Rome: Istituto Gregorio Mendel, 1955.

LAMY, M., MAROTEAUX, P., and BADER, J. P.: Étude génétique du gargoylisme. J. Genet. Hum. **6,** 156–78, 1957.

LAMY, M., ROYER, P., and FRÉZAL, J.: Maladies héréditaires du métabolisme chez l'enfant. Paris: Masson, 1958.

LANGE, J.: Die essentielle Hämochromatose. Deutsch. Med. Wschr. **83,** 2035–38, 1958.

LARSON, C. A.: An estimate of the frequency of phenylketonuria in South Sweden. Folia Hered. Pathol. **4,** 40–46, 1954.

LEHMANN, H.: Variations in human haemoglobin synthesis and factors governing their inheritance. Brit. Med. Bull. **15,** 40–46, 1959.

LEHMANN, H., and SIMMONS, P. H.: Sensitivity to suxamethonium. Apnoea in two brothers. Lancet **2,** 981–82, 1958.

LEJEUNE, J.: Sur une solution "a priori" de la méthode "a posteriori" de Haldane. Biometrics **14,** 513–20, 1958.

LEJEUNE, J., GAUTIER, M., and TURPIN, R.: Les chromosomes humaines en culture de tissus. C. R. Acad. Sci. (Paris) **248,** 602, 1721, 1959.

LEJEUNE, J., TURPIN, R., and GAUTIER, M.: Le mongolisme. Premier exemple d'aberration autosomique humaine. Ann. Genet. **1**, 41–49, 1959.

LELONG, M., BORNICHE, KREISLER, and BAUDY: Mongolien issu de mère mongolienne. Arch. Franc. Pediat. **6**, 231, 1949.

LENNOX, BERNARD: Chromosomes for beginners. Lancet **1**, 1047, 1961.

LENZ, F.: Die Bedeutung der statistisch ermittelten Belastung mit Blutsverwandtschaft der Eltern. Münch. Med. Wschr. **66**, 1340–42, 1919.

———: Methoden der menschlichen Erblichkeitsforschung. *In:* E. GOTSCHLICH: Handb. d. hygien. Untersuchungsmethoden. Jena: G. Fischer, 1929.

———: *In:* E. BAUER, E. FISCHER, and F. LENZ: Menschliche Erblehre. 4th ed. Munich: J. F. Lehmann, 1936.

———: Über kombinantes Verhalten alleler Gene. Erbarzt **5**, 83, 1938.

LENZ, W.: Rotgrün-Blindheit bei einem heterogametischen Schein-Mädchen, zugleich ein Beitrag zur Genetik der heterogametischen Pseudofemininität. Acta Genet. Med. (Rome) **6**, 231–346, 1957.

———: Der Einfluss des Alters der Eltern und der Geburtennummer auf angeborene pathologische Zustände beim Kind II. Acta Genet. **9**, 169–201, 249–93, 1959.

———: Genetisch bedingte Störungen der embryonalen Geschlechtsdifferenzierung. Deutsch. Med. Wschr. **85**, 268–74, 1960.

LENZ, W., NOWAKOWSKI, H., PRADER, A., and SCHIRREN, C.: Die Ätiologie des Klinefelter-Syndroms. Ein Beitrag zur Chromosomenpathologie beim Menschen. Schweiz. Med. Wschr. **89**, 727, 1959.

LEUCHTENBERGER, C., LEUCHTENBERGER, R., and DAVIS, A. M.: A microphotometric study of the desoxyribose nucleic acid (DNA) content in cells of normal and malignant human tissues. Amer. J. Path. **30**, 65–85, 1954.

LEUCHTENBERGER, C., VENDRELY, R., and VENDRELY, C.: A comparison of the content of desoxyribose nucleic acid (DNA) in isolated animal nuclei by cytochemical and chemical methods. Proc. Nat. Acad. Sci. USA **37**, 33–38, 1951.

LEVAN, A.: Self-perpetuating ring chromosomes in two human tumours. Hereditas **42**, 366–72, 1956.

LEVINE, P., ROBINSON, E., CELANO, M., BRIGGS, O., and FALKINBURG, L.: Gene interaction resulting in suppression of blood group substance B. Blood **10**, 1100–1108, 1955.

LEVIT, S. G.: The problem of dominance in man. J. Genet. **33**, 411, 1936.

LEWIS, E. B.: Pseudoallelism and gene evolution. Cold Spring Harbor Symp. on Quant. Biol. **16**, 159–74, 1951.

LIEBENAM, L.: Beitrag zum familiären Vorkommen von Spalthänden und Spaltfüssen. Z. Menschl. Vererb. Konstitutionsl. **22**, 138–51, 1938.

LINDEGÅRD, B.: Variations in human body-build. A somatometric and X-ray cephalometric investigation on Scandinavian adults. Acta Psychiat. Scand., Suppl. 86, 1953.

LIPTON, E. L.: Elliptocytosis with hemolytic anemia: the effects of splenectomy. Pediatrics **15**, 67–83, 1955.

LOEFFLER, L.: Röntgenschädigungen der männlichen Keimzelle und Nachkommenschaft. Ergebnisse einer Umfrage bei Röntgenärzten und -technikern. Strahlentherapie **34**, 735–66, 1929.

LYON, M. F.: Sex chromatin and gene action in the mammalian X-chromosome. Amer. J. Human Genet. **14**, 135–48, 1962.

MACHT, S. H., and LAWRENCE, P. S.: National survey of congenital malformations resulting from exposure to roentgen radiation. Amer. J. Roentgenol. **73,** 442–46, 1955.

MCILROY, J. H.: Hereditary ptosis with epicanthus. A case with pedigree extending over four generations. Proc. Roy. Soc. Med. **23,** 285–88, 1930.

MCINTOSH, R.: The problem of congenital malformation. J. Chron. Dis. **10,** 139–51, 1959.

MACKAY, R. P., and MYRIANTHOPOULOS, N. C.: Multiple sclerosis in twins and their relatives. Preliminary report on a genetic and clinical study. Arch. Neurol. Psychiat. **80,** 667–74, 1958.

MCKENZIE, H. J., and PENROSE, L. S.: Two pedigrees of ectrodactyly. Ann. Eugen. (Lond.) **16,** 88, 1951–52.

MCKUSIK, V. A.: Vererbbare Störungen des Bindegewebes. Stuttgart: G. Thieme, 1959.

MCLEAN, N., and OGILVIE, R. F.: Quantitative estimation of the pancreatic islet tissue in diabetic subjects. Diabetes **4,** 367, 1955.

MALBRAIN, H., PONLOT, R., and DENYS, P.: L'hypophosphatasie et ses manifestations osseuses. Arch. Franc. Pediat. **14,** 337–59, 1957.

MARKS, P. A., GROSS, R. T., and HURWITZ, R. E.: Gene action in erythrocyte deficiency of glucose-6-phosphate dehydrogenase: tissue enzyme levels. Nature (Lond.) **183,** 1266–67, 1959.

MAROTEAUX, P., and LAMY, M.: La dysplasie polyépiphysaire. Sem. Hôp. (Paris) **35,** 3155–69, 1959a.

———: La maladie de Morquio. *Ibid.* **35,** 3147–55, 1959b.

———: L'achondroplasie. *Ibid.* **35,** 3450–63, 1959c.

MARQUARDT, H.: Natürliche und künstliche Erbänderungen. Probleme der Mutationsforschung. Hamburg: Rowohlt, 1957.

MARX, R.: Über Haemophilien und Pseudohaemophilien. Münch. Med. Wschr. **101,** 881–87, 1959.

MATHER, K.: Biometrical genetics. London: Methuen, 1949.

MAU, H.: Wesen und Bedeutung der enchondralen Dysostosen. Stuttgart: G. Thieme, 1958.

MIALL, W. E., and OLDHAM, P. D.: A study of arterial pressure and its inheritance in a sample of the general population. Clin. Sci. **14,** 459, 1955.

MILCH, R. A.: Studies in alcaptonuria: Inheritance of 47 cases in eight highly interrelated Dominican kindreds. Am. J. Hum. Genet. **12,** 76–85, 1960.

MITSCHRICH, H.: Zwillingstuberkulose III. Nachuntersuchung nach 20 Jahren an der Serie tuberkulöser Zwillinge von K. Diehl und O. v. Verschuer. Stuttgart: G. Fischer, 1956.

MOHR, O. L., and WRIEDT, C.: A new type of hereditary brachyphalangy in man. Carnegie Institution of Washington Publication No. 295, 1919.

MONTALENTI, G.: The genetics of microcythemia. Cardiologia, Suppl. 6, pp. 554–58, 1954.

MØRCH, E. T.: Chondrodystrophic dwarfs in Denmark. Op. ex domo biol. hered. hum. Univ. Hafniensis (Copenhagen), No. 3, 1941.

MORTIMER, E. A.: Familial constitutional precocious puberty in a boy three years of age. Pediatrics **13,** 174–77, 1954.

MORTON, N. E.: The detection and estimation of linkage between the genes for elliptocytosis and the Rh blood type. Amer. J. Hum. Genet. **8,** 80–96, 1956.

MORTON, N. E.: Further scoring types in sequential tests, with a critical review of autosomal and partial sex linkage in man. *Ibid.* **9,** 55–75, 1957.

MOTULSKY, A. G.: Metabolic polymorphisms and the role of infectious diseases in human evolution. Hum. Biol., **32,** 28–62, 1960.

MÜHLMANN, W. E.: Geschichte der Anthropologie. Bonn: Universitäts-Verlag, 1948.

MÜLLER, W. A.: Die Biochemie der Erbfaktoren. Münch. Med. Wschr. **97,** 1208–11, 1955.

MÜNTZING, ARNE: Vererbungslehre. Stuttgart: G. Fischer, 1958.

MULLER, H. J.: Radiation damage to the genetic material. *In:* S. A. BAITSELL (ed.), Science in progress, 7th ser., pp. 93–177. New Haven: Yale University Press, 1951.

MURAYAMA, M., and INGRAM, V. M.: Comparison of normal adult human haemoglobin with haemoglobin I by "Fingerprinting." Nature (Lond.) **183,** 1959.

MURRAY, J. E., MERRILL, J. P., and HARRISON, J. H.: Kidney transplantation between seven pairs of identical twins. Ann. Surg. **148,** 343, 1958.

NACHTSHEIM, H.: Strahlengenetik bei Säugetier und Mensch. Bundesgesundheitsblatt, pp. 217–23, 1959.

NEEL, J. V.: A study of major congenital defects in Japanese infants. Am. J. Hum. Genet. **10,** 398–445, 1958.

NEEL, J. V., and SCHULL, W. J.: The effect of exposure to the atomic bombs on pregnancy termination in Hiroshima and Nagasaki. Nat. Acad. Sci.–Nat. Res. Council Publ. No. 461. Washington, D.C., 1956.

NEWMAN, H. H., FREEMAN, F. N., and HOLZINGER, K. J.: Twins, a study of heredity and environment. Chicago: University of Chicago Press, 1937.

NILSSON, I. M., BERGMAN, S., REITALU, J., and WALDENSTRÖM, J.: Haemophilia A in a "girl" with male sex-chromatin pattern. Lancet **2,** 264–66, 1959.

NILSSON, I. M., BLOMBÄCK, M., THILÉN, A., and FRANCKEN, I. VON: Carriers of hemophilia A. A laboratory study. Acta Med. Scand. **165,** 357–70, 1959.

NISHIMURA, E. T., HAMILTON, H. B., KOBARA, TH. Y., TAKAHARA, SH., OGURA, Y., and DOI, K.: Carrier state in human acatalasemia. Science **130,** 333–34, 1959.

Nobel Prizewinners (lead article), Lancet **2,** 655–56, 1959.

NOWAKOWSKI, H., LENZ, W., and PARADA, J.: Diskrepanz zwischen Chromatinbefund und genetischem Geschlecht beim Klinefelter-Syndrom. Acta Endocr. **30,** 296–320, 1959.

NÜRNBERGER, L.: Können Strahlenschädigungen der Keimdrüsen zur Entstehung einer kranken oder minderwertigen Nachkommenschaft führen? Fortschr. Röntgenstr. **27,** 369, 1919.

NYLANDER, E. S.: Präaxiale Polydaktylie in fünf Generationen einer schwedischen Sippe. Upsala Läkarefören. Förhandl. N.F. **36,** 275–92, 1931.

OLDFELT, V.: Incontinentia pigmenti. J. Pediat. **54,** 335–48, 1959.

OLDHAM, P. D., PICKERING, G., FRASER ROBERTS, J. A., and SOWRY, G. S. C.: The nature of essential hypertension. Lancet **1,** 1085–93, 1960.

OUTHIT, M. C.: A study of the resemblance of parents and children in general intelligence. Arch. Psychol. **149,** 1933.

OWEN, R. D.: Genetic aspects of tissue transplantation and tolerance. Am. J. Hum. Genet. **11,** 366–83, 1959.

OWENS, N., HALEY, R. C., and KLOEPFER, H. W.: Hereditary blepharophimosis, ptosis and epicanthus inversus. J. Int. Coll. Surg. **33,** 558–74, 1960.

PARNELL, R. W.: Behaviour and physique. London: Edward Arnold, 1958.

PATAU, K., SMITH, D. W., THERMAN, E., INHORN, S. L., and WAGNER, H. P.: Multiple congenital anomaly caused by an extra autosome. Lancet **1**, 790–93, 1960.

PEARSON, K., and LEE, A.: On the laws of inheritance in man. 1. Inheritance of physical characters. Biometrika **2**, 357, 1903.

PENROSE, L. S.: The problem of anticipation in pedigrees of dystrophia myotonica. Ann. Eugen. (Lond.) **14**, 125–32, 1947–49.

———: The genetical background of common diseases. Acta Genet. **4**, 257–65, 1953.

———: Parental age and mutation. Lancet **2**, 312–13, 1955.

———: Mutation in man. Acta Genet. **6**, 169–82, 1956.

———: The mutational origin of hereditary eye diseases. Mod. Probl. Ophthal. **1**, 501–11. Basel–New York: S. Karger, 1957.

———: Parental age in achondroplasia and mongolism. Amer. J. Hum. Genet. **9**, 167–69, 1957.

PENROSE, L. S., ELLIS, J. R., and DELANTY, J.: Chromosomal translocations in mongolism and in normal relatives. Lancet **2**, 409–10, 1960.

PENROSE, L. S., and STERN, C.: Reconsideration of the Lambert pedigree (ichthyosis hystrix gravior). Ann. Hum. Genet. **22**, 258–83, 1958.

PERKOFF, G. T., NUGENT, C. A., DOLOWITZ, D. A., STEPHENS, F. E., CARNES W. H., and TYLER, F. H.: A follow-up study of hereditary chronic nephritis. Arch. Intern. Med. **102**, 733–46, 1958.

PFÄNDLER, U.: La manifestation hétérozygote et homozygote de certains troubles du métabolisme (porphyrie chronique, cystinose, maladie de Niemann-Pick). J. Genet. Hum. **5**, 248–60, 1956.

PFÄNDLER, U., and BERGER, H.: Cystinose (Cystinspeicherkrankheit) und ihre Beziehungen zur Cystinurie und Hyperaminoacidurie. Ann. Paediat. **187**, 1–41, 1956.

PHILIP, U., and WALTON, J. N.: Colour blindness and the Duchenne-type muscular dystrophy. Ann. Hum. Genet. **21**, 155–58, 1956.

PICKERING, G. W.: High blood pressure. London: Churchill, 1955.

PICKFORD, R. W.: Colourblindness and its inheritance. Biol. and Hum. Affairs **21**, 19–26, 1956.

———: Some heterozygous manifestations of colour blindness. Brit. J. Physiol. Opt. **16**, 83–95, 1959.

PIPER, J., and ORRILD, L.: Essential familial hypercholesterolemia and xanthomatosis. Amer. J. Med. **21**, 34–46, 1956.

PLATT, R.: The nature of essential hypertension. Lancet **2**, 55–57, 1959.

PONTECORVO, G.: Trends in genetic analysis. London: Oxford University Press, 1959.

PORTER, H. M.: Immunologic studies in congenital agammaglobulinemia with emphasis on delayed hypersensitivity. Pediatrics **20**, 958–65, 1957.

POULSON, D. F.: The effects of certain X-chromosome deficiencies on the embryonic development of Drosophila melanogaster. J. Exp. Zool. **83**, 271–326, 1940.

PRADER, A.: Die Häufigkeit des kongenitalen adrenogenitalen Syndroms. Helv. Paediat. Acta **13**, 426–31, 1958.

PRADER, A., and SIEBENMANN, R. E.: Nebenniereninsuffizienz bei kongenitaler Lipoidhyperplasie der Nebennieren. Helv. Paediat. Acta **12**, 569–95, 1957.

PRANKERD, T. A. J., ALTMAN, K. I., and YOUNG, L. E.: Abnormalities of carbohydrate metabolism of red cells in hereditary spherocytosis. J. Clin. Invest. **33**, 957, 1954.

QUICK, A. J., and HUSSEY, C. V.: Hemophilia B (PTC deficiency, or Christmas disease). Arch. Intern. Med. **103**, 762–75, 1959.

RACE, R. R., and SANGER, R.: Blood groups in man. 3d ed. Oxford: Blackwell Scientific Publications, 1958.

———: Inheritance of blood groups. Brit. Med. Bull. **15**, 99–108, 1959.

RATH, B.: Rotgrünblindheit in der Calmbacher Blutersippe. Nachweis des Faktorenaustausches beim Menschen. Arch. Rassenbiol. **32**, 397–407, 1938.

RATNER, B., and SILBERMAN, D. E.: Allergy—its distribution and the hereditary concept. Ann. Allergy **10**, 1–20, 1952.

RAYNER, S.: Juvenile amaurotic idiocy. Diagnosis of heterozygotes. Acta Genet. **3**, 1–5, 1952.

REHN, A. T., and THOMAS, E. J.: Familial history of a mongoloid girl who bore a mongoloid child. Amer. J. Ment. Defic. **62**, 496–99, 1957.

RENWICK, J. H.: Nail-patella syndrome: evidence for modification by alleles at the main locus. Ann. Hum. Genet. **21**, 159–69, 1956.

RENWICK, J. H., and LAWLER, S. D.: Genetical linkage between the ABO and nail-patella loci. Ann. Hum. Genet. **19**, 312–31, 1955.

RIEGER, R., and MICHAELIS, A.: Genetisches und cytogenetisches Wörterbuch. 2d ed. Berlin-Göttingen-Heidelberg: Springer, 1958.

ROOS, J., and HUIZINGA, J.: Genetic investigation of the Stuart coagulation defect. Acta Genet. **9**, 115–22, 1959.

ROSENOER, V. M., and FRANGLEN, G.: Caeruloplasmin in Wilson's Disease. Lancet **2**, 1163–64, 1959.

RUNNER, M. N.: Inheritance of susceptibility to congenital deformity. Embryonic instability. J. Nat. Cancer Inst. **15**, 637–49, 1954.

RUSSELL, W. L., RUSSELL, L. B., and CUPP, M. B.: Dependence of mutation frequency on radiation dose rate in female mice. Proc. Nat. Acad. Sci. USA **45**, 18–23, 1959.

RUSSELL, W. L., RUSSELL, L. B., and KELLY, E. M.: Radiation dose rate and mutation frequency. The frequency of radiation induced mutation is not, as the classical view holds, independent of dose rate. Science **128**, 1546–50, 1958.

SARKAR, S. S., BANERJEE, A. R., BHATTACHARJEE, P., and STERN, C.: A contribution to the genetics of hypertrichosis of the ear rims. Amer. J. Human Genet. **13**, 214–23, 1961.

SCHETTLER, G., JOBST, H., KÄPPLER, H. P., and KÖRFGEN, G.: Familiäre Hypercholesterinämie. Deutsch. Med. Wschr. **84**, 368–74, 356–57, 1959.

SCHLEGEL, W.: Körper und Seele. Eine Konstitutionslehre für Ärzte, Juristen, Pädagogen und Theologen. Stuttgart: F. Enke, 1957.

SCHULL, W. J.: The effect of Christianity on consanguinity in Nagasaki. Amer. Anthrop. **55**, 74–88, 1953.

———: Empirical risks in consanguineous marriages: sex ratio, malformation, and viability. Am. J. Hum. Genet. **10**, 294–343, 1958.

SCHULL, W. J., and NEEL, J. V.: Radiation and the sex ratio in man. Science **128**, 343–48, 1958.

SCHULZE, CHR.: Erbbedingte Strukturanomalien menschlicher Zähne. Munich and
    Berlin: Urban & Schwarzenberg, 1956.
SCHWARTZ, E. E., and UPTON, A. C.: Factors influencing the incidence of leukemia,
    special consideration of the role of ionizing radiation. Blood 13, 845–64,
    1958.
SCOTT, E. M., and HOSKINS, D. D.: Methemoglobinemia in Alaskan Eskimos and
    Indians. Blood 13, 795–802, 1958.
SEIP, M.: Lipodystrophy and gigantism with associated endocrine manifestations.
    A new diencephalic syndrome? Acta Paediat. 48, 555–74, 1959.
SENDI, H.: Quelques cas de Keratosis follicularis spinulosa decalvans (Siemens)
    Thèse. Geneva, 1957.
SHEPPARD, P. M.: Blood groups and natural selection. Brit. Med. Bull. 15, 134–
    39, 1959.
SIMPSON, C. L., and HEMPELMANN, L. H.: The association of tumors and roent-
    gen-ray treatment of the thorax in infancy. Cancer 10, 42–56, 1957.
SJÖGREN, T.: Genetic-statistical and psychiatric investigations of a West Swedish
    population. Acta Psychiat. Scand., Suppl. 52, pp. 1–102, 1948.
SLATIS, H. M., REIS, R. H., and HOENE, R. E.: Consanguineous marriages in the
    Chicago region. Am. J. Hum. Genet. 10, 446–64, 1958.
SLOME, D.: The genetic basis of amaurotic family idiocy. J. Genet. 27, 363–76,
    1933.
SMITH, C. A. B.: A note on the effects of method of ascertainment on segregation
    ratios. Ann. Hum. Genet. 23, 311–23, 1959.
SMITH, C. A. B., and KILPATRICK, S. J.: Estimates of the sex ratio mutation rates
    in sex-linked conditions by the method of maximum likelihood. Ann. Hum.
    Genet. 22, 244–49, 1958.
SMOLLER, M., and HSIA, D. Y.-Y.: Studies on the genetic mechanism of cystic
    fibrosis of the pancreas. J. Dis. Childr. 98, 277–92, 1959.
SMYTH, C. J.: Hereditary factors in gout. Metabolism 6, 218–29, 1957.
SMYTH, C. J., COTTERMAN, C. W., and FREYBERG, R. H.: The genetics of gout and
    hyperuricemia—analysis of nineteen families. J. Clin. Invest. 27, 749–59, 1958.
SNYDER, L. H., and DOAN, C. A.: Is the homozygous form of multiple teleangi-
    ectasia lethal? J. Lab. Clin. Med. 29, 1211–16, 1944.
SØBYE, P.: Heredity in essential hypertension and nephrosclerosis. Op. ex domo
    biol. hered. hum. Univ. Hafniensis (Copenhagen), No. 16, 1948.
SØRENSEN, H. R.: Hypospadias, with special reference to aetiology. Copenhagen:
    E. Munksgaard, 1953.
SORSBY, A. (ed.): Clinical genetics. London: Butterworth, 1953.
SPIEGELMANN, M., and MARKS, H. H.: Age and sex variations in the prevalence
    and onset of diabetes mellitus. Amer. J. Public Health 1, 26, 1946.
SPINDLER, E. A.: Über die Häufigkeit von Verwandtenehen in drei württember-
    gischen Dörfern. Arch. Rassenbiol. 14, 9–12, 1922.
SPURLING, C. L., and SACKS, M. S.: Inherited hemorrhagic disorder with anti-
    hemophilic globulin deficiency and prolonged bleeding time (vascular hemo-
    philia). New Engl. J. Med. 261, 311–19, 1959.
STANBURY, J. B., MEIJER, J. W. A., and KASSENAAR, A. A. H.: The metabolism of
    iodo-tyrosines: II. The metabolism of mono- and di-iodotyrosines in certain
    patients with familial goiter. J. Clin. Endocr. 16, 848, 1956.
STEPHENS, F. E., and TYLER, F. H.: Studies in disorders of muscle: inheritance of

childhood progressive muscular dystrophy in thirty-three kindreds. Amer. J. Hum. Genet. **3**, 111, 1951.

STERN, C.: Grundlagen der menschlichen Erblehre. Göttingen-Berlin-Frankfurt: Musterschmidt, 1955.

———: The problem of complete Y-linkage in man. Amer. J. Hum. Genet. **9**, 147–65, 1957.

———: Le daltonisme lié au chromosome X a-t-il une localisation unique ou double? Exposé de deux théories. J. Genet. Hum. **7**, 302–7, 1958.

STERN, C., and WALLS, G. L.: The Cunier pedigree of "Colour Blindness." Am. J. Hum. Genet. **9**, 249–73, 1957.

STEVENSON, A. C.: Achondroplasia: account of the condition in Northern Ireland. Amer. J. Hum. Genet. **9**, 81–91, 1957.

STEVENSON, A. C., and CHEESEMAN, E. A.: Hereditary deaf mutism, with particular reference to Northern Ireland. Ann. Hum. Genet. **20**, 177–231, 1956.

STEVENSON, A. C., and MARTIN, V. A. F.: Retinoblastoma. Occurrence of the condition in Northern Ireland, 1938–1956. Brit. J. Prev. Soc. Med. **11**, 29–35, 1957.

STRÖMGREN, E.: Beiträge zur psychiatrischen Erblehre. Copenhagen, 1938.

SUTTER, J.: Recherches sur les effects de la consanguinité chez l'homme. Biol. Med. (Paris) **47**, 563–660, 1958.

SVERDRUP, A.: Postaxial polydactylism in six generations of a Norwegian family. J. Genet. **12**, 217–40, 1922.

TABOR, E. C., and FRANKHAUSER, K. H.: Detection of diabetes in a nutritional survey—a study of 550 persons in Ottawa County, Michigan. Public Health Rep. **65**, 1330–35, 1950.

TANNER, J. M.: Inheritance of morphological and physiological traits. *In:* A. SORSBY (ed.): Clinical genetics. London: Butterworth, 1953.

———: The genetics of human morphological characters. Advancement of Science **51**, No. 13, 192–94, 1956.

THUMS, K.: Eineiige Zwillinge mit konkordanter multipler Sklerose. Wien. Z. Nervenheilk. **4**, 173–203, 1951.

TORREGROSA, M. V. DE, ORTIZ, A., and VARGAS, D.: Sickle cell-spherocytosis associated with hemolytic anemia. Blood **11**, 260–65, 1956.

TOUSTER, O.: Pentose metabolism and pentosuria. Amer. J. Med. **26**, 724–39, 1959.

TREVOR-ROPER, P. D.: Marriage of two complete albinos with normally pigmented offspring. Brit. J. Ophthal. **36**, 107–8, 1952.

TURPIN, R., LAFOURCADE, J., CHIGOT, P. L., and SALMON, C.: Présomption de monozygotisme en dépit d'un dimorphisme sexuel: sujet masculin XY et sujet neutre Haplo X. C. R. Acad. Sci. (Paris) **252**, 2945–46, 1961.

TURPIN, R., LEJEUNE, J., LAFOURCADE, J., and GAUTIER, M.: Aberrations chromosomiques et maladies humaines. La polydyspondylie à 45 chromosomes. C. R. Acad. Sci. (Paris) **248**, 3636–38, 1959.

USHER, C. H.: A pedigree of epicanthus and ptosis. Ann. Eugen. (Lond.) **1**, 128–38, 1925–26.

———: On a few hereditary eye affections. Trans. Ophthal. Soc. UK **55**, 164–245, 1935.

VAN ARSDEL, P., and MOTULSKY, A. G.: Frequency and heredity of asthma and allergic rhinitis in college students. Acta Genet. **9**, 101–14, 1959.

VANDEMBRONCKE, J., VERSTRAETE, M., and VERWILGHEN, R.: L'afibrinogénémie congénitale. Présentation d'un nouveau cas et revue de la littérature. Acta Haemat. **12,** 87–105, 1954.

VANDEPITTE, J., and DELAISSE, J.: Sicklémie et paludisme. Aperçu du problème et contribution personelle. An. Soc. Belge Med. Trop. **37,** 703, 1958.

VANDEPITTE, J. M., ZUELZER, W. W., NEEL, J. V., and COLAERT, J.: Evidence concerning the inadequacy of mutation as an explanation of the frequency of the sickle cell gene in the Belgian Congo. Blood **10,** 341–50, 1955.

VAN GEFFEL, R., DEVRIENDT, A., DUSTIN, J. P., URIS, H., and LOEB, H.: La maladie du galactose. Considerations génétiques. Étude de l'amino-acidémie et de l'amino-acidurie. Arch. Franc. Pediat. **15,** 158–84, 1959.

VENDRELY, R., and VENDRELY, C.: L'acide desoxyribonucléique (DNA). Substance fondamentale de la cellule vivante. Paris: Amédée Legrand, 1957.

VERSCHUER, O. VON: Der erste Nachweis von Faktorenaustausch (crossing over) beim Menschen. Erbarzt **5,** 3, 1938.

———: Genetik des Menschen. Lehrbuch der Humangenetik. Munich and Berlin: Urban & Schwarzenberg, 1959.

VERSCHUER, O. VON, and KOBER, E.: Die Frage der erblichen Disposition zum Krebs. Ergebnis einer Forschung durch 20 Jahre an einer auslesefreien Zwillingsserie. Akad. Wiss. Abhandl. Math. nat. Kl. Jg. No. 4. Wiesbaden: Franz Steiner, 1956.

VIGNES, ———.: Epicanthus héréditaire. Recueil d'Ophthal. **11,** 422–25, 1889.

VOGEL, F.: Vergleichende Betrachtungen über die Mutationsrate der geschlechtsgebundenen-rezessiven Hämophilieformen in Dänemark und der Schweiz. Blut **1,** 91, 1955.

———: Über die Prüfung von Modellvorstellungen zur spontanen Mutabilität an menschlichem Material. Z. Menschl. Vererb. Konstitutionsl. **33,** 470–91, 1956.

———: Modellvorstellungen zur spontanen Mutabilität beim Menschen. Berliner Med. **8,** 96–99, 1957*a*.

———: Neue Untersuchungen zur Genetik des Retinoblastoms (Glioma retinae). Z. Menschl. Vererb. Konstitutionsl. **34,** 205–36, 1957*b*.

———: Gedanken über den Mechanismus einiger spontaner Mutationen beim Menschen. *Ibid.* **34,** 389–99, 1958*a*.

———: Zur Problematik induzierter Mutationen beim Menschen. Röntgen-Blätter **11,** 113–205, 1958*b*.

———: Moderne Anschauungen über Aufbau und Wirkung der Gene. Deutsch. Med. Wschr. **84,** 1825–33, 1959*a*.

———: Moderne Probleme der Humangenetik. Ergebn. Inn. Med. Kinderheilk. **12,** 52–125, 1959*b*.

VOGEL, F., and WENDT, G. G.: Zwillingsuntersuchung über die Erblichkeit einiger anthropologischer Masse und Konstitutionsindices. Z. Menschl. Vererb. Konstitutionsl. **33,** 425–46, 1956.

WAARDENBURG, P. J.: Eine Reihe erblich-angeborener familiärer Augenmissbildungen. Graefe's Arch. Ophthal. **124,** 221–29, 1930.

WALD, N.: Leukemia in Hiroshima City atomic bomb survivors. Science **127,** 699–700, 1958.

WALLS, G. L.: Peculiar color blindness in peculiar people. Arch. Ophthal. **62,** 13–37, 1959.

WALTON, J. N.: On the inheritance of muscular dystrophy. Ann. Hum. Genet. **20,** 1–38, 1955.

WEBER, ERNA: Grundriss der biologischen Statistik für Naturwissenschaftler, Landwirte und Mediziner. 3d ed. Jena: G. Fischer, 1957.

WEDLER, A. W., and WELSCH, A.: Über ein erbliches Missbildungssyndrom mit Beckenhörnern ("Turnersches Syndrom"). Z. Menschl. Vererb. Konstitutionsl. 31, 243–53, 1952.

WEINBERG, W.: Methoden und Fehlerquellen der Untersuchung auf Mendelsche Zahlen beim Menschen. Arch. Rassen Gesellschaftsbiol. 9, 165–74, 1912.

WEINER, R. L., and FALLS, H. F.: Intermediate sex-linked retinitis pigmentosa. Arch. Ophthal. 53, 530–35, 1955.

WEISMANN, A.: Das Keimplasma. Eine Theorie der Vererbung. Jena: G. Fischer, 1892.

WEITZ, W.: Zur Ätiologie der genuinen oder vasculären Hypertension. Z. Klin. Med. 96, 151–81, 1923.

WIENER, A. S.: Rh-Hr blood types. New York: Grune & Stratton, 1954.

WILDERVANCK, L. S.: Hereditary congenital anomalies of bones and nails in five generations. Luxation of the capitulum radii, luxation or absence, resp. hypoplasia of the patella, crooked little fingers and dystrophy or absence of the nails and abnormal lunulae. Genetica 25, 1–28, 1950.

WILLEBRAND, E. A. VON, and JÜRGENS, R.: Über ein neues vererbbares Blutungsübel: Die konstitutionelle Thrombopathie. Deutsch. Arch. klin. Med. 175, 453–83, 1933.

WINGFIELD, A. H.: Twins and orphans. The inheritance of intelligence. London, Dent, 1928.

WINTERS, R. W., GRAHAM, J. B., WILLIAMS, T. F., McFALLS, V. M., and BURNETT, CH. H.: Genetic study of familial hypophosphatemia and vitamin D resistant rickets with a review of the literature. Medicine 37, 97–142, 1958.

WISKOTT, A.: Familiärer, angeborener Morbus Werlhofii. Mschr. Kinderheilk. 68, 212–16, 1937.

WITKOP-OOSTENRIJK, G. A.: Contribution to the study of the inheritance of dysostosis cleidocranialis. Acta Genet. 7, 223, 1957.

WOLFF, J. A., and BERTUCIO, M.: A sex-linked, genetic syndrome in a Negro family manifested by thrombocytopenia, eczema, bloody diarrhea, recurrent infection, anemia and epistaxis. Amer. J. Dis. Child. 93, 74, 1957.

WORK, T. S.: Protein biosynthesis: some connecting links between genetics and biochemistry. Proc. Growth Symposium, ed. R. O. ERICKSON. Madison, Wis., 1959.

WULZ, G.: Ein Beitrag zur Statistik der Verwandtenehen. Arch. Rassenbiol. 17, 82–95, 1925.

ZACHAU, H. G.: Die Biosynthese der Nukleinsäuren. Zur Verleihung des Nobelpreises für Medizin 1959 an A. Kornberg und S. Ochoa. Deutsch. med. Wschr. 85, 230–33, 1960.

INDEX